ANTIMONY, GOLD, AND JUPITER'S WOLF

How the
elements
were named

ANTIMONY
GOLD AND
JUPITER'S
WOLF

PETER
WOTHERS

OXFORD
UNIVERSITY PRESS

OXFORD
UNIVERSITY PRESS

Great Clarendon Street, Oxford, OX2 6DP,
United Kingdom

Oxford University Press is a department of the University of Oxford.
It furthers the University's objective of excellence in research, scholarship,
and education by publishing worldwide. Oxford is a registered trade mark of
Oxford University Press in the UK and in certain other countries

First Edition published in 2019

Impression: 1

Published in the United States of America by Oxford University Press
198 Madison Avenue, New York, NY 10016, United States of America

British Library Cataloguing in Publication Data
Data available

Library of Congress Control Number: 2019945159

ISBN 978–0–19–965272–3

Printed and bound in Great Britain by
Clays Ltd, Elcograf S.p.A.

I dedicate this book to
Joanne and Keith Rutherford
who not only share their name with
the discoverer of the most abundant element
in the atmosphere, but who, in the course of their valuable professions,
save lives by administering the second most abundant element,
just as foreseen by its discoverer
250 years ago.

PREFACE

In November 2016, the International Union of Pure and Applied Chemistry (IUPAC) officially announced the proposed names for four new elements—nihonium, moscovium, tennessine, and oganesson. These four additions finally 'completed' the periodic table, in that every space had a named element from number 1 (hydrogen) to number 118 (oganesson), and all seven rows of the table were now filled. It's possible that other elements might be synthesized—research is certainly ongoing—but it's unlikely that the table will ever appear as neat again, since to fill the next row, another fifty-four elements would need to be made.

The naming process is not a trivial matter, and it usually takes several years. An independent body (a joint working party between the IUPAC and the International Union of Pure and Applied Physics) first has to scrutinize the evidence and confirm that atoms of the element in question were actually formed (albeit briefly), then establish who exactly made them first, and, finally, wait for all the concerned parties within the group who first made the discovery to agree on a name. Even then, the joint working party has to approve the suggested name. Despite the arduous process, it's clear how these four elements were named: nihonium after one of the two ways to say 'Japan' in Japanese and literally meaning 'the Land of Rising Sun'; moscovium after the Moscow region, home of the Joint Institute of Nuclear Research, where the experiments resulting in its discovery were conducted; tennessine in recognition of the contribution of the Tennessee region of the United States to superheavy research, and, finally, oganesson, named after one of the pioneers of transactinoid element research, Professor Yuri Oganessian.

It's easy to guess where the names of some of the other elements came from—for example, einsteinium or germanium. But for many others, their origins are less clear or even not known with certainty. These are the elements of interest in this book. It isn't a simple etymological list of the elements and their names; rather, it's an exploration of why some of the elements eventually ended up with the names they did. For example, while it's easy to find out that selenium was named after the goddess of the moon, why did the discoverer choose to do that in the first place? The answer to this question is much more involved. Similarly, the element oxygen is familiar to all of us, but it has a far-from-ideal name, supposedly meaning 'acid-generator', although others from the period thought it more precisely suggested 'a sharp chin', or even 'the son of a vinegar merchant'. Exactly how it ended up with the name requires a tour through the history of chemistry and an understanding of some of the key scientists and their changing theories.

Where possible, original sources have been used so that we can hear the language as used by the original discoverers (or, at least, an English translation). Old spellings have been retained, with the exception of the 'long s' character or the occasional v/u interchange.

ACKNOWLEDGEMENTS

First, I must thank my parents for giving me my very first chemistry set and believing me when I promised not to 'blow the house up'. There was, I think, only one 'minor' disaster, when a jar of the intolerably smelly gas hydrogen sulfide toppled over and stank the house out. I can only apologize again, and I am sure it would not have helped at the time if I had pointed out that the gas was more toxic than hydrogen cyanide...I must also thank my brother John and his wife Tracy for their constant love and support; and my sister Joanne and her husband Keith, who are acknowledged in the dedication. I am grateful to my whole family for the encouragement and support that led to the first generation from our family attending university.

From the home I grew up in, to what has now been home for most of my life: I must acknowledge the invaluable support I receive from the community that is St Catharine's College, Cambridge. This has come in the form of moral support, guidance, and encouragement from the Master of the College, currently in the form of Sir Mark and Lady Welland, and previously from past Masters Dame Jean Thomas, Prof. David Ingram, and Sir Terrance English. I have been inspired by some great fellow scientists—notably the late Sir Alan Battersby, a highly distinguished chemist, and also the late Dr John Shakeshaft, who admitted a terrified young teenager to the college many years ago. Our community thrives with a melting pot of great minds of different hues (rather like the elements of the periodic table); it is a privilege, for example, to end up sitting next a leading expert on Shakespeare at lunch, or to have coffee with a medieval French linguist. I thank many of my fellow Fellows for their help with parts of this project, in particular Dr Paul Hartle (English), Dr Miranda Griffin (medieval French), Prof. Geoffrey Kantaris (Spanish), Drs John and Dorothy Thomson (History and Classics respectively), Dr Hester Lees-Jeffries (English), Prof. Richard Dance (Anglo-Saxon Norse and Celtic), and Dr River Chen and Prof. Hans van de Ven for their much-needed help as I tried to understand the meaning behind Chinese characters. I am also grateful to other members of our Fellowship who have played a key role, directly or indirectly, in the production of this book, namely Prof. John Pyle (Chemistry), Drs Ivan and Helen Scales (Geography and Marine

Biology respectively), Dr John Little (Materials), Dr Jess Gwyne (Materials), Mr Simon Summers (Bursar), Mrs Deborah Loveluck (Development Director), and former Fellows now at other universities: Prof. Jim MacElwaine (Maths), Prof. John Gair (Maths), and Prof. Paul Raithby (Chemistry).

I must also thank my colleague, friend, and previous co-author, Dr James Keeler of Selwyn College. He has always been someone I (like many others) can only aspire to emulate with his superb clarity of mind, great knowledge of the subject of chemistry, and much appreciated teaching skill. I have learned much from James, as have thousands of students. Also from our department, I must acknowledge the wonderful Emma Powney, who manages to organize my life and somehow ensures that I end up lecturing at the right venue at the right time; Emma's comments on the early drafts of this work, so many years ago, were invaluable.

Of course, our institution exists to educate students, and I have had the honour of playing a small part in this over a number of years. I hope other educators also relish the pleasure of encountering a mind sharper than one's own (although this does not, fortunately, occur *every* day). Many students have studied Natural Sciences at St Catharine's, and I am proud to have a number of my former students as close friends now. While I feel reluctant to name any, I must give thanks to James Brimlow, Claire Badger, and Jed Kaniewski for their support and friendship over the years, and more recently to current PhD students Peter Bolgar, who went through the entire draft of this book and came back with many helpful suggestions; George Trenins, who cheerfully helped out with translations from Russian; and Andrea Chlebikova, with her keen interest in the history of chemistry, and her fantastic computer skills.

While his student days did not overlap with my time at St Catharine's, one of our former Natural Scientists, Peter Dawson, and his wife Christina must both be thanked for the incredible support they have given and continue to give to the College, but particularly, from the perspective of this work, for the amazing gift of material relating to the early history of the periodic table. To help us celebrate the 150th anniversary of Mendeleev's first periodic table in 2019, they have generously helped to establish St Catharine's College as holders of one of the finest collections anywhere outside of Russia of works from the great chemist, including the very first printings of the periodic table in its different forms. More than this though, they have become very close friends who have eagerly awaited the appearance of this book.

On the subject of rare books, almost all of the references quoted in this book, together with the images reproduced, are taken from my own library. Finally,

I can justify why I 'needed' those twenty different editions of Lavoisier's *Traité*! The collection has amassed over more than thirty years, and while these have come from many friendly dealers from all over the world, I must particularly acknowledge the support, advice, and encouragement (!) of Graham Weiner, Roger Gaskell, Nigel Phillips, Jonathan Hill, Julien Comellas, James Burmester, and Hugues de Latude.

Of those who have watched, bewildered, as this library grew from my days at school, I cannot omit my friends Jon Cardy, Duncan Curtayne, Adam Loveday, Colin Tregenza Dancer, Stephen Driver, and my chemistry teacher David Berry, with whom I still regularly discuss the subject. Similarly, for putting up with my many chemical stories over the years, thanks to Tim Hersey, Andrew and Laura Worrall, Kathryn Scott, Ben Pilgrim, Penny Robotham, and Andy Taylor. For general support and sanity checks, but also help with translations from German, I thank Susanna Prankel—ably assisted by Kat and Leo. For further help with translations from German (which seemed to be the hardest language to translate properly from early works), I am grateful to Manfred Kerschbaumer and the much-missed, late Wolfgang Hampe.

From Oxford University Press, I must thank for this stray into 'trade books', Latha Menon and Jenny Nugée, who have endured the slow gestation of this book over too many years while continuing to encourage and guide me. I should also acknowledge those who oversaw my first publishing adventures, Michael Rodgers and Jonathan Crowe. The anonymous reviewer needs to be thanked for their wise suggestions and heartening words, but I am also grateful for the valuable discussions I have had with other experts in the field, Hasok Chang, Jenny Rampling, Frank James, and Charlotte New. Special thanks to my cousin, Stuart Clayton, who inspired the title for Chapter 4 after a late-night discussion, many years ago.

Finally, I must thank three people very dear to me. First of all, my partner Umut Dursun. In addition to the constant encouragement over the years that have helped keep the project going, Umut has also been understanding and supportive in dealing with my obsession with rare chemistry texts. Except in a few cases where indicated, all of the images featured in the book have been photographed and edited by Umut from our collection, and for this and so much more, I am truly thankful. The last two people to be mentioned are Christopher Lamb and Gordon Hargreaves—two very dear friends who have become close family to us.

CONTENTS

NOTE ON THE ILLUSTRATIONS

All figures aside from Figures 24, 31, 36, and 43 are from the author's private collection. Figure 45 is courtesy of the Master and Fellows of St Catharine's College, Cambridge.

The naming of a new element is no easy matter.
For there are only twenty-six letters in our alphabet,
and there are already over seventy elements.

—William Ramsay, 'An Undiscovered Gas', 1897.

One hundred and eighteen elements are known today.

1

HEAVENLY BODIES

Lift up thy head and looke upon the heaven,
And I will learne thee truly to know the Planets seaven.
—Ashmole, 1652[1]

We don't know for sure where the names of the longest-known elements come from, but a connection was made early on between the most ancient metals and bodies visible in the heavens. Figure 1 shows an engraving from a seventeenth-century text with the title 'The Seven Metals' (translated from the Latin). It isn't immediately obvious how the image is meant to depict seven metals until we explore the connections between alchemy and astronomy. However strange such associations seem to us now, we shall see that new elements named in the eighteenth, nineteenth, twentieth, and twenty-first centuries have had astronomical origins. We can't properly understand why some of the more recent elements were named as they were without first understanding these earlier historical connections.

The Seven Wanderers

As we look into the night sky, the distant stars remain in their same relative positions and seem to move gracefully together through the heavens. Of course, we now know that it is the spinning Earth that gives this illusion of movement. The imaginations of our ancestors joined the bright dots to pick out fanciful patterns such as the Dragon, the Dolphin, or the Great Bear—the latter being more often known today (with rather less imagination) as the Big Dipper, the Plough, or even the Big Saucepan. But, while these patterns, the constellations, remained unchanging over time, there were seven objects, or 'heavenly bodies', that seemed to move across the skies with a life of their own. They were given

Fig. 1. This engraving, taken from the *Viridarium Chymicum* of Daniel Stolcius from 1624, depicts 'The Seven Metals'.

the name 'planet', which derives from the Greek word for 'wanderer' ('*planetes asteres*', '*πλάνητες ἀστέρες*', meaning 'wandering stars').

These seven bodies were the Sun, the Moon, Mercury, Venus, Mars, Jupiter, and Saturn, all of which were documented by the Babylonians over three thousand years ago. Until the sixteenth century, the most commonly held view was that the Earth was at the centre of the Universe and that the seven bodies revolved around the Earth, with the relative orbits shown schematically in Figure 2. The order is nicely described in the early seventeenth century by the Polish alchemist Michael Sendivogius in his book *A New Light of Alchymie*: 'Thou seest that *Saturne* is placed the uppermost, or highest, next to *Iupiter*, then *Mars*, then *Sol*, or the Sun, then *Venus*, then *Mercury*, and last of all *Luna*, or the Moon.'[2]

Now, of course, we would classify only five of the original seven as planets, which, together with the Earth, orbit the Sun. The Moon orbits around our Earth just as we orbit the Sun; this is shown schematically in Figure 3.

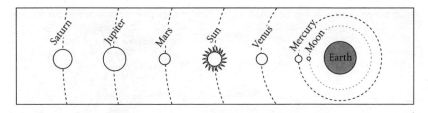

Fig. 2. The geocentric view of the solar system, as understood by the earliest astronomers.

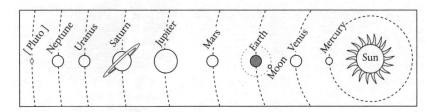

Fig. 3. The modern heliocentric view of the solar system.

The modern, heliocentric view of our solar system was first formulated by Nicolaus Copernicus (1473–1543) in the early part of the sixteenth century. So profound was his theory that it is often taken to mark the beginning of the modern astronomy that led the scientific revolution. In 2009 his achievements were recognized with a place in the chemists' periodic table; man-made element 112 was officially named copernicium, bringing the relationship between astronomy and chemistry into the twenty-first century.

The remaining planets in the solar system could not be discovered until the invention of suitable telescopes. Then Uranus was observed in 1781; Neptune in 1846; and, finally, in 1930, Pluto, which was a planet until its reclassification in 2006 to the status of 'dwarf planet'. But we are leaping ahead of ourselves; we'll return to these most distant planets later. First, we must look at the original seven heavenly bodies.

The number seven has long been held to have a certain mystical significance. There are seven days of the week reflecting the seven days of creation in the Bible. The Seven Deadly Sins are balanced by the Seven Heavenly Virtues. In Islam there are seven levels in heaven and the same number in hell. Rome was founded upon Seven Hills. It has been said that Isaac Newton divided the rainbow into seven colours in order to imitate the seven notes in a musical scale.

Over time, each of the seven heavenly bodies came to be associated with a particular day of the week and with one of the gods from ancient mythology—Sunday

The bodyes ſeuyn loke hem here anon	The seven bodies view them here anon
Sol gold is /& luna ſiluer we threpe	Sol Gold, and Luna we maintain is Silver;
Mars Iren Mercury quykſiluer we clepe	Mars is iron, Mercury we call quicksilver
Saturnus led and Iubiter is tyn	Saturn is lead, and Jupiter is tin
And venus copyr by my fader kyn	And Venus copper on my family's honour.

Fig. 4. The 'seven bodies' from Chaucer, reproduced in the black letter typeface as used by Caxton. The modern interpretation is provided courtesy of scholar of seventeenth-century English Paul Hartle.

The bodyes whiche I ſpeke of here	The bodies I speak of here
Of the planettes ben bygonne	Originate from the planets
The gold is tytled to the ſonne	The gold is assigned to the sun
The mone of ſyluer hath his part	The moon to silver gives his place
And Iren that ſtond vpon mart	And Iron stands for Mars
The leed after Saturne growweth	The lead takes growth from Saturn
And Inbiter the bras beſtowweth	And Jupiter bestows the brass
The Coper ſette is to Venus	The copper derives from Venus
And to his part Mercurius	And for his part Mercury
Hath the quyk ſyluer as it falleth	Takes in order the quicksilver

Fig. 5. The 'seven bodies' from Gower, as printed by Caxton in the fifteenth century, together with a modern interpretation by Paul Hartle.

with the Sun, Monday with the Moon, Tuesday ('*mardi*' in French, where the link is more obvious) with Mars, Wednesday ('*mercredi*') with Mercury, Thursday ('*jeudi*') with Jupiter, Friday ('*vendredi*') with Venus, and Saturday with Saturn. Surely it could be no coincidence that there were also just seven metals known in ancient times—gold, silver, copper, iron, tin, lead, and mercury. Each of these also became associated with a particular heavenly body. While the specific associations varied a little over time, a general consensus was reached around the fifteenth century. This is nicely exemplified in Geoffrey Chaucer's *Canterbury Tales*, written in the fourteenth century, and beautifully printed by William Caxton in the fifteenth. The relevant extract, together with a modern interpretation, is shown in Figure 4.

A similar verse appears in the *Confessio Amantis* by John Gower, a contemporary of Chaucer. This work was also printed by Caxton in the fifteenth century, and is shown in Figure 5.

One might consider how this idea of a connection between the planets and the metals developed. It was common knowledge that the rays from the sun are necessary to nurture growing plants. God did not make plants fully formed, but created seeds that were then nourished by sunlight and rain. The same idea was extended to metals, as Sendivogius writes:

And what prerogative should Vegetables have before Metalls, that God should put a seed into them, and without cause withhold it from these? Are not Metalls of as much esteem with God as Trees? Let this be granted for a truth, that nothing grows without seed: for where there is no seed, the thing is dead. It is necessary therefore that four Elements should make the seed of Metalls...hee which gives no credit to this undoubted truth, is not worthy to search into the secrets of nature...[3]

So it was also believed by many that metals and minerals 'grew' from seed in the ground. There is a certain logic to this: many elements and minerals do form only in particular regions within the earth. But the process usually takes thousands or millions of years, and it does not involve the creation of the element from seed, but the concentration of what already exists into one particular form in one location. John Webster in his *Metallographia, or, A History of Metals*, published in 1671, adopts quite a modern view when he opposes the miner's belief that metals are formed ready-made by God in the ground and instead states that before the metals penetrate into rocks, cliffs, and stones they were in a solution 'either in form of water, or vapours, and steams'.[4]

Although he did not personally believe in the connections between the planets and the metals, the French chemist Nicolas Lemery gives an excellent account of what some believed in his *Course of Chymistry* from 1677:

Astrologers have conceited that there was so great an affinity and correspondence between the Seven *Metals* before named, and the seven *Planets*, that nothing happened to the one, but the others shared in it; they made this correspondence to happen through an infinite number of little bodies that pass to and from each of them; and they suppose these corpuscles to be so figured that they can easily pass through the pores of the *Planet* and *Metal* they represent, but cannot enter into other bodies because their pores are not figured properly to receive them; or else if they do chance to get admittance into other bodies, they can't fix and stay there to contribute any nourishment; for they do imagine that the *Metal* is *nourished* and perfected by the *Influence* that comes from its *Planet*, and so the *Planet* again the same from the *Metal*.

For these reasons they have given these seven *Metals* the name of the seven *Planets*, each accordingly as they are governed: and so have called *Gold* the *Sun*, *Sylver* the *Moon, Iron Mars, Quicksilver Mercury, Tinn Jupiter, Copper Venus*, and *Lead Saturn*.

They have likewise fancied that each of these *Planets* has his day apart to distribute liberally his Influence on our Hemisphere; and so they tell us that if we work upon *Sylver* on *Munday, Iron* on *Tuesday*, and so of the rest, we shall attain our end much better than on other daies.

Again they have taught us that the seven Planets do every one govern some particular principal part of our bodies; and because the *Metals* do represent the *Planets*, they must needs be mighty *specifick* in curing the distempers of those parts, and keeping them in good plight. Thus they have assigned the *Heart* to *Gold*, the *Head* to *Sylver*, the *Liver* to *Iron*, the *Lungs* to *Tinn*, the *Reins* [kidneys] to *Copper*, and the *Spleen* to *Lead*.[5]

Lemery adds that this is only what the 'most sober' astrologers say, and that the theories of others are even more absurd. He goes on to state:

'Tis no hard matter to disprove these conceits, and shew how groundless they are; for no body ever yet got near enough to the *Planets*, to satisfie himself whether they are really of the same nature with *Metals*, or whither any *Effluviums* of bodies do fall from them to us.

Lemery concludes that many excellent remedies may be prepared from the metals, but that the effects of these preparations 'may better be explicated by Causes nearer at hand than the *Stars*'.

Let's consider each of these associations between the metal and planet in turn, together with some of the 'excellent *Remedies*' that may be drawn from them.

The Sun and Gold

Figure 6 shows representations of gold taken from alchemical texts of the seventeenth century, in which the metal is referred to by its astronomical counterpart. The obvious connection between the Sun and metallic gold is the brilliant colour of each. The Sun is unique in our solar system; indeed, it is the defining body. Gold metal also has properties that make it quite distinct from the other known metals. Most metals change in their appearance over time—iron rusts; a white crust covers lead; silver tarnishes black; copper turns green; and so on. In contrast, gold is chemically inert; it does not react with anything in the air or the ground, and so retains its brilliance seemingly forever. Golden artefacts dug up after centuries in the earth look as glorious as the day they were fashioned. This chemical resilience extends to its reaction with acids since gold is the only metal that is inert to the common mineral acids. For instance, the other six metals readily dissolve in what we now call nitric acid. This acid was known in earlier times as *aqua fortis*, or 'strong water', because of this very ability to dissolve solids. However, something stronger still is required

SOL. Gold. AURUM : Metallorum Rex.

SOLEIL.

Fig. 6. Solar references to gold in chemical texts. The illustration on the left (1662) shows gold as the king of the metals, and it includes the alchemical symbol for gold in the mouth of the lion. The woodcut on the right (1673) shows a goldsmith at work, hammering the metal to make gold leaf.

to dissolve gold: a special mix of two acids, nitric and hydrochloric. This king of all acids is known as *aqua regia*, or 'royal water'.

The chemical symbol for gold used by chemists today is Au, derived from the Latin *aurum*, but the alchemists used a circle—the perfect geometrical figure—as their symbol to represent both gold and the Sun. Gold was thought to be the perfect metal, and all the other metals were thought to gradually mature in the ground until they ultimately reached the perfection of gold. Webster, in his book *Metallographia*, states: 'Natures ultimate labour is in time to bring all Metals to the perfection of Gold: which she would accomplish, if they were not unripe and untimely taken forth of the bowels of the Earth.'[6] One of the goals of the alchemist was to use artful manipulation to speed up this natural process whereby the imperfect metals are gradually matured into gold.

The association between the Sun and gold is neatly summarized in Christopher Glaser's *The Compleat Chymist* from 1677: '…it is justly called the King of Metals, as being the most perfect of all. 'Tis called also the Sun, as well for the resemblance it hath with the Sun of the great World which enlightens us; as for that it hath with the heart of man, which is the Sun of our little World.'[7]

The connection between the Sun and the heart is just one of the many influences that the planets were thought to hold over various parts of our bodies. Indeed, the very word 'influenza' is thought to arise from such unfavourable astral influences. Such fanciful connections were promoted by the London doctor William Salmon in the second half of the seventeenth century. He suggested, rather specifically, that the Sun 'rules the Heart, Arteries, Back and Sight, Right Eye of a Man, and the Left of a Woman'.[8] Perhaps not surprisingly, when it came to diseases, Glaser adds that gold 'signifies all Passions of the Heart; as Faintings, Tremblings, Swoundings, Pimples in the Face, Red Choler, Weakness of the Sight, Burning Fevers, putrid and rotten'.

In other seventeenth-century chemistry books and dispensatories we find such recipes as 'Solar Diaphoretick' (to promote perspiration)[9] and 'aurum fulminans' (used for 'diseases proceeding from corruption of the Blood').[10] This latter preparation, also known as 'thundering gold'.[11] It's actually a highly explosive compound prepared by adding ammonia to a solution of gold dissolved in aqua regia. Lemery in his *Course of Chymistry* from 1686 explains the origin of the explosion when the compound is put in the fire. He says 'the great Detonation, or noise that it makes, cannot proceed from any thing else, but the inclosed Spirits which violently divide the most compact body of Gold to get out quickly, when they are forced to it by the action of Fire'.[12]

Lemery reassures us that 'we need not fear lest *Aurum Fulminans* taken inwardly, and heated by the stomach, should cause such a *Detonation* there, as it does when set over the fire in a spoon; for so much the more moisture as comes to it, so much the less noise does it make'. Curiously, modern medicine still continues to prescribe high explosives to treat diseases of the heart; in small doses, nitroglycerine, the explosive agent in dynamite, is a most effective treatment for angina.

The historical links between the Sun, the Roman god Sol, and the metal gold are now virtually forgotten, but a new element keeps the connections alive—the element helium. As we shall see in Chapter 8, this element is deserving of a name linking it to the Sun since it was first detected there in 1868. The only element to have been first discovered off our planet, helium is named after the Greek personification of the Sun, Helios. When it was finally isolated on Earth, it was found not to be a metal but instead an incredibly unreactive gas. To this day, helium remains the only non-metallic element to have the suffix -*ium*, which is otherwise reserved for metals such as sodium, chromium, and uranium.

Silver and the Moon

As with gold and the Sun, there is an obvious connection between the appearance of the Moon and metallic silver (Figure 7). Glaser writes: '*Silver* is a Metal less fixed, less weighty, and less perfect than *Gold*, though much more so than all other Metals, and passes for a perfect Metal, because it comes near the perfection of *Gold*. 'Tis called *Luna* from its colour, and from the great Remedies it affords for the diseases of the Brain, which by sympathy easily receives the impressions of the *Moon*.'[13]

The connection between the moon and diseases of the brain remains with us today in the word '*lunatic*', originally associated with insanity that was thought to recur with the phases of the moon. Salmon extends the influences: 'she rules the Bulk of the Brain, the Stomach, Bowels, Bladder, Left Eye of a Man, Right of a Woman'.[14] Preparations of silver were 'held a special strengthener of the Brain, to comfort the Animal Spirits; good in all Head diseases, Epilepsies, Apoplexyes &c'.[15]

LUNA. Silber. ARGENTUM.

LVNE.

Fig. 7. Silver in chemical texts represented by the Moon. On the left, the goddess Luna, representing silver, features the alchemical symbol for the metal on her head. On the right, the silversmith makes chalices and other religious artefacts.

The use of the astronomical name in medicinal silver preparations continued throughout the nineteenth century and into the twentieth; 'lunar caustic', now known as silver nitrate, was commonly used to cauterize wounds.

While the alchemical symbol for silver was the crescent, for modern chemists it's Ag. This derives from the Latin for silver, *argentum*—which also gives its name to Argentina, the only country that can be said with certainty to have been named after an element. Although a number of countries have elements named after them (France, Poland, Germany, and America), the reverse is less common. It is possible that there is one other instance—Cyprus—but we will come to that later.

Mercury—Element, God, and Planet

Mercury is the only one of the ancient metals still to share its name with both the planet and the god (Figure 8). As with the rest of the ancient metals, its connection to the heavenly body is somewhat obscure. In order to understand these relationships, we need to recall some basic astronomy.

As the planet closest to the Sun, Mercury is hard to see since it never strays far from the solar glare. For this reason, it was probably the last of the planets

MERCURIUS. ARGENTUM VIVUM. Quecfilber.

Fig. 8. Seventeenth-century illustrations representing the element mercury: on the left, personified as the god holding his wand or caduceus, and on the right, being used as an emetic.

to be discovered by the ancients. To the nearest whole number, the Earth takes 365 days—one year—to move once round the Sun. However, the time taken to complete one orbit depends on how close the orbiting body is to the Sun. Mercury takes just 88 (Earth) days to complete one cycle, and Venus, the next closest, takes 225 days. In contrast, Saturn, the farthest known planet of the ancient world, takes close to 29½ Earth years to complete one revolution around the Sun. Consequently Saturn is seen to move through space very slowly relative to the fixed patterns of the constellations, while Mercury, when visible, positively zips along.

This fast motion, nipping back and forth to the Sun, was no doubt part of the reason that the planet Mercury was associated with the messenger of the gods in Roman mythology. Being fleet of foot, the messenger god is easily recognizable, with wings on his heels and helmet. The element itself is no slouch—it is the only metal that is liquid at room temperature, not solidifying until −39° C. In Old English, mercury is usually known as quicksilver (with varying spellings). The 'quick' here, rather than indicating speed, means living or alive; this usage prevails in the expression 'the quick and the dead', and in the 'quick' of our fingernails. The original Latin for metallic mercury was *argentum vivum*—literally 'living silver'. Our modern symbol for mercury, Hg, comes from *hydrargyrum*, a modernized Latin version of the Greek *hydrargyros*, meaning 'water silver'.

The alchemical symbol used for mercury derived from the staff or wand carried by the god, his caduceus (Figure 9). Occasionally this has mistakenly been used as a symbol by medical practices (for example, by the US Army Medical Corps). The god Mercury, or Hermes, had nothing to do with the practice of medicine; this was carried out by another god in Greek mythology, Asclepius. His symbol was a staff entwined by a single snake, rather than the double snake of Hermes.

One of mercury's strange properties is its ability to form amalgams with other metals. When solid gold is added to mercury, the gold dissolves to form gold amalgam. This reaction has been used as a way of purifying gold since the mercury will only react with the gold, leaving any stony matrix of rock behind. Heating the amalgam then vaporizes the mercury so that only pure gold remains. Mercury amalgamates with most metals; the few exceptions include iron, which floats on top of a pool of mercury, and platinum, which, being denser, readily sinks unchanged to the bottom. Mercury is prohibited on aeroplanes because any spills could have catastrophic consequences, steadily dissolving the aluminium-based hull. Amalgams are still widely used today to make fillings in dentistry: when a solid alloy of silver, tin, and copper is mixed

Mercury and Quickſilver.

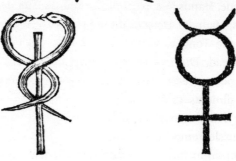

Fig. 9. The alchemical symbol (right) for both the planet and metal is the wand or caduceus of the god Mercury.

with liquid mercury, the whole mass softens and can be easily moulded into a cavity, where it hardens after a couple of minutes.

It is this ability of mercury to mix with other metals that gave rise to another connection between the planet and the element. Since Mercury is the fastest-moving planet, it has a greater chance of appearing in the same region of space as some of the other planets when viewed from Earth. Such an occurrence is known in astronomy (and astrology) as a conjunction. This explains the seventeenth-century description of mercury from Glaser: 'Quick-silver is a Mineral body fluid, heavy and shining…'Tis called *Mercury* from its conformity in its Actions with the *Celestial Mercury*, which frequently mixes its influences with those of other Planets, and according to its different Conjunction produces different Effects: so our *Mercury* easily joyns with other Metals, and diversifies its effects according to the quality which it gives or receives from the Metallick Bodies and Mineral Spirits with which it is joyned.'[16]

The use of mercury in medicine has always been controversial, with its reputation alternating between that of a universal medicine and a deadly poison. The 1730s saw the publication of *A Treatise upon the Use and Properties of Quicksilver* by 'A Gentleman of Trinity College Cambridge' (taken to be a certain Dr Dover). In the preface, Dover states: 'FORMERLY People were afraid to take it; the common Apprehensions of its dire Effects, its poisonous Qualities, had rendered it a Terror to all. NOW, strange Alteration! These Fears are fled; its taken in every little Disorder, by Children too, without Disguise, without Mixture, crude and in Substance; and 'tis as usual to meet with it in Families, as Snuff or Tobacco.'[17] A few years later appeared *Mercury Stark Naked* by Isaac Swainson, which aimed to 'strip that poisonous mineral of its medical pretentions'.

A century earlier, Salmon had written that Mercury 'rules the Brain, Imagination, Tongue, Hands and Feet: *Of Diseases*, such as are incident to the Brain; as Vertigoes, Madness, Defects of the Memory, Convulsions, Asthma's, Imperfections of the Tongue, Hoarsness, Coughs, Snuffling in the Nose, Stopages in the Head, Dumbness, and whatsoever hurts the Intellectual Faculty'.[18] What is not entirely clear at this point is whether mercury is thought to help cure such ailments, or induce them in the first place. Certainly by the nineteenth century, the symptoms of mercury poisoning—including the development of muscle tremors and changes in the mental faculties and behaviour—were recognized in various professions. A well-known example is the hat-making business, in which the compound mercury nitrate was routinely used to stiffen rabbit fur. Frequent exposure gave rise to the condition known as 'hatters' shakes' and the expression 'as mad as a hatter'; such cases probably influenced Lewis Carroll in his portrayal of the hatter in *Alice in Wonderland*.

At one time, metallic mercury was even prescribed for use as a laxative. Far from being disagreeable, Dover states that it is 'inviting, it looks like a finer Jelly, is tasteless on the Tongue, and hardly felt in going down, or in the Stomach'.[19] Mercury prescriptions continued well into the twentieth century, even as its toxicity was becoming much better understood.

Iron and Mars

Mars was the Roman god of war; his Greek counterpart was Ares. It is not too much of a leap to connect the metal used in making the instruments of war to the god and the red planet (Figure 10). The alchemical symbol for both the planet and metal—which is also used by biologists to indicate 'male'—is thought to derive from the shield and spear of the martial god (Figure 11). The symbol used by modern chemists is Fe, from *ferrum*, the Latin for iron. This root has also given rise to the term 'farrier' (older version: 'ferrer') for a blacksmith.

In his *Royal Pharmacopoea* of 1678, Moyse Charas writes, 'Iron is call'd by the name of *Mars*, whether employed for the making of weapons of war, of which *Mars* was said to be the God; or because of the influences which Iron receives from that Planet.'[20] Others noted that the appearance of the red planet bore a resemblance to the red-hot coals of the blacksmith, or indeed to red-hot iron itself. As it turns out, there are connections between the planet and the metal that were not known to the ancients. The red colour of Mars is due to the presence of iron oxides on the planet's surface. These oxides are commonly

Fig. 10. Images from the seventeenth century representing the element iron. On the left, iron is represented by Mars, the Roman god of war. On the right, we see a cutler or knife-grinder sharpening iron tools.

Mars and Iron.

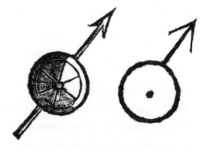

Fig. 11. The alchemical symbol for iron as derived from the weapons of the god of war.

found on Earth as the mineral haematite, which in turn gets its name from its blood-red colour. The mineral is described in John Maplet's 1567 book *A Greene Forest*: 'Ematites is a stone somewhat ruddie, somewhat sanguine, found both in Affrick, in Indie and in Arabie: so named for that it resolveth & chaungeth oft into a bloudie colour: and is called of some stench bloud, for that it stoppeth his vent or course of flowing.'[21]

The use of haematite for stopping the flow of blood might have been suggested by the appearance of the weathered mineral, which resembles a large blood clot; but we now know that iron is an essential element for the production of blood in the body. The molecule haemoglobin, which transports oxygen around our bodies in red blood cells, contains at its very heart an atom of iron. It is this atom that binds reversibly to oxygen from the air, and less reversibly, and with lethal consequences, to the poison gas carbon monoxide. Since a lack of iron in our diets leads to anaemia, the element is often added as a supplement to food—for instance, some breakfast cereals contain powdered metallic iron, which readily dissolves in the acid of our stomachs. Given how simple and effective this treatment is, it is curious that iron was not recognized as a treatment for anaemia until the nineteenth century. Certainly many preparations containing iron were used, such as vitriol of Mars (iron sulfate), good for 'obstructions of the *Liver, Spleen, Pancreas*, and *Mesentry*'; an 'Aperitive Tincture of *Mars* by means of *tartar*', which was 'also very good against the Worms and putrifaction of the Stomach and Bowles'; and the 'Crocus or Safron of Mars', thought to be good for gonorrhoea. But it seems none of these treatments were prescribed specifically for blood-related diseases.[22] Salmon states that Mars 'rules the Gall, left Ear, the Apprehension and Smell, and the Bulk of the Head and Face'; he doesn't mention the blood at all.

The female counterpart of Mars was the goddess Venus, associated with the element copper. We will come to her shortly, but to put her in the proper context, we first need to explore some more mythology.

Lead and Saturn

Saturn is the planet furthest from the Earth still visible to the naked eye. It is also the slowest-moving planet and with the largest orbit, and it may well be this that gave rise to its connection with the dense, 'ponderous' metal lead. In early literature, both lead and the god Saturn are often represented by a lame old man (Figure 12), again referring to the slow-moving characteristic of the planet.

According to Salmon, Saturn 'rules the Spleen, right Ear, Bones, Teeth Joynts'. Lead is now known to be poisonous—it accumulates in the bones and prevents the incorporation of iron in haemoglobin. Despite this, plenty of medicinal preparations of Saturn were used well into the early nineteenth century. One common substance was Magisterium Saturnium, also known as 'sugar of lead'

SATURNUS. Bley. PLUMBUM.

Fig. 12. Images from the seventeenth century representing the element lead. The god Saturn, left, is usually depicted with a scythe, which has been suggested as the origin of the alchemical symbol for lead. On the right, a leadsmith pours out the molten metal to make sheets used for lining roofs. Lead piping was used to carry water from Roman times until well into the twentieth century.

and now known to chemistry students as lead(II) ethanoate or lead acetate. This was prepared by heating up lead in air to form lead oxide, and then adding vinegar. The acetic acid (ethanoic acid) in the vinegar dissolves the oxide, thereby forming the lead acetate. It was referred to as a sugar since it tastes remarkably sweet and was even used to sweeten wine, notably in Roman times (bear in mind that it is highly toxic). It has been suggested that Beethoven died from lead poisoning, possibly from artificially sweetened wine. A 1669 source tells us, 'This sugar inwardly taken, by its coldness, doth also extinguish Venereal Lust; and is therefore profitable for those who are devoted to a single and Virgin life.'[23]

In the late eighteenth century, French physician Thomas Goulard wrote a book titled *A treatise on the effects and various preparations of lead, particularly of the extract of Saturn, for different chirurgical disorders*. By this period, most prescriptions were recommended only for external complaints such as bruises and burns. Goulard gives various testimonies for its successful application, providing a glimpse into life in eighteenth-century France: 'A PAGE of the Duke of Richelieu had an inflamed testicle, owing to a bruise he had received in riding. Many remedies had been prescribed without success, and the disorder

Fig. 13. Lead represented by Saturn in *Chymica Vannus* (1666).

continued to gain ground. Upon seeing him, I ordered compresses of the Vegeto-mineral Water [lead acetate] to be applied to the part; which soon gave him ease. The day after, the pains were entirely abated, and in eight or ten days the cure was compleated.'[24]

The depiction of Saturn from *Chymica Vannus* (Figure 13), published in 1666, is full of symbolism from ancient mythology. The god is shown carrying a scythe and devouring a baby. Similarly gruesome scenes have been depicted by artists such as Peter Paul Rubens in 1636, and Francisco Goya around 1820. The Greek counterpart of Saturn was Kronos (Κρόνος) or Cronus, who is easily confused with a different god, Chronos (Χρόνος), the god of time. Our image of Old Father Time, and perhaps the Grim Reaper, is mainly derived from this latter god—time being the one thing that catches up with each of us in the end—but the scythe motif probably comes from confusion with Saturn/Kronos.

The scythe or sickle (Figure 14) has been suggested as the origin of the alchemical symbol for both the planet Saturn and the metal lead. The modern

Saturn and Lead.

Fig. 14. The scythe of Saturn, giving rise to the symbol for lead.

symbol used by chemists is Pb from the Latin for lead, *plumbum*, from which we still get the terms 'plumb line' (a lead weight) and 'plumber' (traditionally a trade involving work with lead water pipes).

Saturn's scythe has no agricultural significance, but a far more gruesome origin. According to Hesiod's *Theogony*, an account of the descent of the Greek gods written in the eighth century BC, the scythe was fashioned from flint by his mother Gaia, goddess of the earth. It was given to Kronos to help him overthrow his father Uranus (or Ouranos), god of the sky. Hesiod describes what happened after the sky god arrived and spread himself over the earth: 'His son reached out from the ambush with his left hand; with his right he took the huge sickle with its long row of sharp teeth and quickly cut off his father's genitals, and flung them behind him to fly where they might.'[25]

We will return to this tale shortly. As for why Saturn is often depicted eating children, it was foretold that Saturn/Kronos would be overthrown by one of his offspring, and in an attempt to prevent this, he cruelly ate them all at birth. His wife, Rhea, 'suffered terrible grief' and 'begged her dear parents, Earth and starry Heaven, to devise a plan so that she could bear her child in secrecy'. Saturn's evil scheme was thwarted by substituting a swaddling-wrapped stone for a child, which he duly swallowed whole without noticing the deception. The baby grew up and, just as prophesized, overthrew his father and became king of the gods. According to some, in a cruel twist of fate, the son then carried out his revenge by castrating his own father, echoing Saturn's earlier act. This rebel son who became king of all the gods was Zeus in Greek mythology, or the Roman god Jupiter.

Tin and Jupiter

Perhaps the fact that Jupiter is the largest planet in our solar system, as well as the brightest after the beautiful Venus, led to the connection with the king of

IUPITER. ℈m. STANNUM.

Fig. 15. Images from the seventeenth century representing the element tin. On the left, Jupiter is shown on his throne; the top of his sceptre is his symbol, perhaps based on a thunderbolt. On the right, a tinsmith prepares various drinking utensils from tin, or an alloy of tin such as pewter.

the gods Zeus or Jupiter (Figure 15). The connection to the metal is less clear; maybe the pairing was simply by default since this was the last of the seven ancient metals to need a planetary connection. However, it has been suggested that the crackling heard when a piece of tin is bent (so-called tin-cry) might have been reminiscent of the thunder of Zeus.

It is possible that the form of a thunderbolt from the king of the gods may have given rise to the symbol used for tin and Jupiter. Or perhaps it was derived from his throne. Another theory, suggested in 1783 by Oxford physician and professor of chemistry Martin Wall, holds that the symbol stems from a connection between the Roman god Jupiter and the Egyptian god Amun, who was often depicted with the head of a ram. Figure 16, illustrating this idea, is taken from Wall's book.[26] The modern symbol Sn comes from the Latin word for tin, '*stannum*', from where we get the word 'stannary', relating to tin.

Salmon relates that Jupiter 'rules the Lungs, Liver, Veins & Blood', and Glaser states of tin, "Tis called Jupiter, by reason of the afinity it hath with the Jupiter of the great World, and for that the Remedies made of it serve for the Diseases

Jupiter and Tin.

Fig. 16. The alchemical symbol for tin, taken from an eighteenth-century dissertation on the origin of the symbols.

Saturnus hyght a kyng of Crete	There was a king of Crete called Saturn
He had be put out of his sete	He needed to be removed from his throne
He was put doune as he whiche stood	He is recorded as one possessed
In frenesye a was so wood	By frenzy, and was so insane
That fro his wyf whiche Rea hyght	That from his wife, named Rhea
His owne children he to plyght	He seized his own children
And ete hem of his comon wone	And made it his usual custom to eat them.
But Jupyter whiche was his sone	But his son Jupiter
And of ful age his fader bonde	Grown to full age, bound his father
And kyt of with his owne hande	And with his own hand severed
His genytalles whiche also fast	His genitals, which at once
In to the depe see he cast	He threw into the depths of the sea
Wherof the grekies afferme a seye	On account of which the Greeks claim and record
Thus when they were cast awey	That when they were scattered
Cam Venus forth by wey of kynde	Venus was born by this means

Fig. 17. Verses from Gower's *Confessio Amantis* as printed by Caxton in the fifteenth century, together with a modern interpretation by Paul Hartle.

of Liver and the Matrix.' Tin remedies do not seem to have been particularly common, although the 'Salt of Jupiter' (probably tin ethanoate) is 'endued with very great virtues in all *Hysterical* Diseases'.[27]

Tin metal was much needed in ancient times since adding it to copper improves the latter's strength, giving the superior alloy bronze. Cornwall, in the southwest of Britain, was known across Europe for its tin production, and the region may have been exporting tin from the beginning of the second millennium BC. It has even been suggested that the metal gave rise to the name for the country: Britain. In his textbook *Elements of Chemistry*, first published in Latin in 1732, Hermann Boerhaave (1668–1738), the Dutch professor of chemistry, botany, and physics at the University of Leyden, writes of tin: 'The best sort of it is found in *Great Britain*, which yields vast quantities; and hence *Bochart* was led to conjecture that the word *Bretania* was derived from the *Syriac Barat*

Anac, which signifies a field of Tin.'[28] Although the seventeenth-century French scholar Samuel Bochart may have proposed this origin for the word 'Britain' (via the contraction 'Bratanac'), there are several other suggestions, including being derived from the word 'brith', signifying the blue dye woad, or simply as meaning 'northern island' from *bor-i-tain*.

Returning to the myths of the ancients, Hesiod tells us that after Saturn emasculated his father he cast the severed flesh into the sea. In the fourteenth century, Gower says in Old English verse that it was Jupiter who threw Saturn's genitals into the sea (Figure 17). This is how the goddess Venus is said to have been formed: from the cast-off flesh of Saturn/Kronos, or, more probably, his father Uranus.

Copper and Venus

Venus was the Roman goddess of love and her Greek counterpart was Aphrodite, which has been interpreted as meaning 'risen from the foam'. Hesiod gives a detailed account of her birth following the castration of Uranus by his son Kronos:

> As for the genitals, just as he first cut them off with his instrument of adamant and threw them from the land into the surging sea, even so they were carried on the waves for a long time. About them a white foam grew from the immortal flesh, and in it a girl formed. First she approached holy Cythera; then from there she came to sea-girt Cyprus. And out stepped a modest and beautiful goddess, and grass began to grow all round beneath her slender feet. Gods and men call her Aphrodite, because she was formed in foam, and Cytherea, because she approached Cythera, and Cyprus-born, because she was born in wave-washed Cyprus, and 'genial', because she appeared out of genitals.[29]

The isle where Aphrodite finally settled, Cyprus (or Kypros, *Κύπρος*) was famed in ancient times for its copper mines. The Latin for 'copper', *cuprum*, possibly derives from the island name, but the reverse is also possible, that the name Cyprus is derived from the name of the metal. Either way, it is from the Latin name that chemists get the modern symbol for copper, Cu (Figure 18).

Given the origin of Venus and the fact that she is the goddess of love, it is hardly surprising that, according to Glaser, the planet 'rules the Womb, Yard [penis], Testicles, all the Instruments of Generation, the Reins [kidneys], Throat, and Womens Breasts'.[30] He states of copper that 'The Chymists call it *Venus*, both by reason of the influences which possibly it receives

VENUS. CUPRUM. Rupffer.

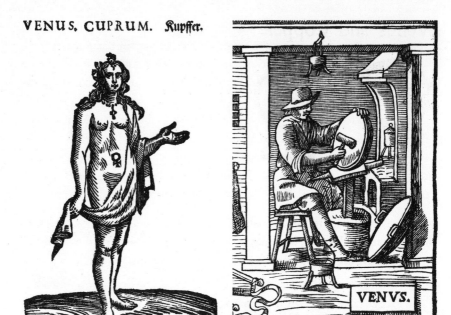

Fig. 18. Images representing the element copper.

from that Planet, and of the virtue it hath in Diseases seated in the parts of Generation.'[31]

It is the connection with these 'diseases in the parts of generation' that gives us the phrase 'venereal disease'. Glaser states that compounds such as 'A Volatile Vitriol and Magistery of *Venus*' (probably copper sulfate) are 'a Soveraign Remedy for an old *Gonorrhea*.'[32] While gonorrhoea is actually a bacterial infection, many similar diseases are fungal. Copper is a most effective antifungal agent and is still used in some parts of the world today to treat fungal infections, particularly on fruit trees. As well as the sulfate, other preparations of copper are sometimes used, such as the ethanoate (or acetate). Copper ethanoate was sometimes referred to as verdigris, but more often this name referred to corroded copper in general. The term is derived from the Old French, literally meaning 'green of Greece'; the colour is familiar to us all from aged copper roofs or, more strikingly, from the famous Statue of Liberty in New York, which started off the colour of metallic copper. Its green patina results from a chemical reaction between copper and oxygen, water, and traces of sulfur-containing gases in the air.

Many early authors suggest that the symbol for copper derives from the looking-glass of Venus. Others, including Martin Wall, have suggested it comes

Venus and Copper.

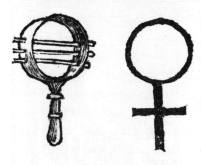

Fig. 19. The alchemical symbol for copper (right), which Wall in 1783 derived from a sistrum (left).

from a sistrum (Figure 19), a musical instrument sometimes associated with the Egyptian goddess Isis, who may have been linked with the goddess Venus.[33] Biologists have appropriated the same symbol for 'female' in general.

The End of the Seven

Not everyone believed in the associations between the planets and metals. As we have already seen, Lemery was most sceptical. But even prior to Lemery, some recognized that any connection probably would not last because new metals and perhaps even new planets would eventually be discovered.

In 1640, the Spanish miner Albaro Alonso Barba wrote:

[T]hey assign the number, names, and colours of the Planets unto Mettals, calling Gold, the Sun; Silver, the Moon; Copper, Venus; Iron, Mars; Lead, Saturn; Quicksilver, Mercury...but this subordination and application is uncertain, as is also the conceit that Mettals are but seven in number; whereas it is very probable, that in the bowels of the Earth there be more sorts than we yet *know*. A few years ago in the mountains of *Sudnos* in *Bohemia*, was found a Mettal which they call *Bissamuto*, which is a Mettal between Tin and Lead, and yet distinct from them both: there are but few that know of it, and 'tis very possible more Mettals also may have escaped the notice of the generality. And if one should admit the subordination, and resemblance between Mettals and the Planets, modern experience, by excellent Teliscopes has discover'd, that they are more than seven. *Gallileo de Galiles* has written a Treatise of the *Satelites* of *Jupiter*, where one may find curious observations of the number and motion of those new Planets.[34]

More elements did indeed come to be known from the fifteenth century, but many of these did not really show the same sort of characteristics as the true metals. Bismuth, antimony, zinc, and cobalt slowly came to be recognized as, if not 'proper' metals, at least semi-metals. However, when platinum was discovered in Peru and its properties studied in the early eighteenth century, there really was no escaping the fact that this truly was a metal, in many ways as good as gold. For a while, platinum was even known as the 'eighth metal'.

Along with new metals came new planets. Although Galileo had observed the four largest moons of Jupiter around 1609, they were not proper planets, i.e. bodies orbiting the Sun. The discovery of a new planet did come about in 1781 with Sir William Herschel's observation of a body he named *Georgium Sidus* (George's Star), in honour of his patron King George III. An alternative name, Uranus, was proposed by the rival astronomer from Germany, Hohann Elert Bode. Bode logically argued that just as Saturn was the father of Jupiter, Uranus was the father of Saturn (excusing the mix of Roman and Greek mythologies). His suggestion name came to be the accepted name (with resistance only from England), and so the new planet took its name from the Greek god of the sky.

In 1786, not long after the discovery of Uranus, German mineralogist Martin Henry Klaproth was studying the properties of a mineral called pitch-blende— the same mineral in which Marie Curie would later discover minute traces of the elements radium and polonium. Klaproth isolated a new metal (actually, it was still the oxide of the metal), which he announced as follows:

> The ancient philosophers, who considered our globe as the center of the material universe; and the sun, on the contrary, merely as a planet destined, like the others, to a periodical circumvolution round the earth, flattered themselves that they had discovered a great mystery of Nature, in the agreement of the *seven celestial bodies*, which they assumed for planets, with the *seven metals* known in those times. In consequence of the various hypotheses which they founded on this supposed mystery, they allotted to each metal a certain planet, by whose astral effluvia its generation and maturation were to be promoted. In like manner, they took from these planets their names and symbols, to designate the metals subordinated to them. But as the above number of metals has long since been increased by later researches; and as the discovery of new planets has not kept pace with that of metals, the metals newly found out have been deprived of the honour of receiving their names from planets, like the older ores. They, therefore, must be satisfied with the name given them accidentally and, in most instances, by the common miner.
>
> Of late, *seventeen* metallic substances have been acknowledged as distinct metals, each of a nature peculiar to itself. The design of this essay is to add one to that number, the chemical properties of which will be explained in the sequel.[35]

Fifteen pages later, after having demonstrated that this is indeed a new element, he finally gets round to naming his discovery. 'I have chosen that of *uranite, (Uranium)*, as a kind of memorial, that the chemical discovery of this new metal happened in the period of the astronomical discovery of the new planet *Uranus*.' In the English edition of this work, there is an asterisk after the planet name Uranus, indicating a footnote from the translator that 'this planet is called *Georgium sidus* in England only'. Curiously, in an earlier publication he proposes to give his new metal the name Uranites (which was quickly altered to the more conventional uranium) only 'until a more suitable one shall be thought of'.[36]

We'll come back to Klaproth shortly, but first we should conclude the discovery of the planets in our solar system. Not long after Herschel's startling discovery, a number of new 'planets' began to pop up. The first of these was discovered on the first of January 1801 when the Italian astronomer Giuseppe Piazzi observed what he first thought was a star—until he noticed it moving against the constellations. Cautiously, he announced it as a comet rather than a planet, but after the famous mathematician Carl Friedrich Gauss correctly predicted its orbit around the Sun between Mars and Jupiter, it was confirmed as a small planet. Piazzi named the new body after the Roman goddess Ceres. On the astronomical scale, Ceres really is very small, with a diameter of no more than 1000 km, about a third that of our moon.

In 1802 another 'planet' was discovered in the same region of space. This one was named Pallas after the Greek goddess of wisdom, Pallas Athena. Then in 1804 yet another was discovered and named Juno, after the Roman goddess and mother of Mars. Eventually these 'minor planets' came to be known as asteroids, after it was realized that there are many even smaller bodies in the same region of the solar system—an asteroid belt, lying approximately between the orbits of Mars and Jupiter. However, before this distinction the first two to be discovered, Ceres and Pallas, ended up with elements named after them: cerium and palladium, each of which was discovered shortly after their heavenly counterparts. There was even a 'junonium' for a while, before it was shown to be identical to cerium. A tongue-in-cheek note to the journal *Nature* published on 25 December 1879, after another (as it later turned out, false) discovery of an element, ended: 'Chemists will have to keep as narrow a watch on these minor elements as our astronomers do upon the minor planets, or we shall not know where we are.'

The next true planet in our solar system, Neptune, was recognized as an orbiting planet in 1846, even though Galileo may have observed it over two hundred years earlier in 1612. The planet is named after the brother of Jupiter, Neptune, the god of the sea (his Greek equivalent being Poseidon).

When the Russian chemist Dmitri Mendeleev proposed his famous classification of the elements, the periodic table, in 1869, uranium was the heaviest element he knew of. There are now many elements placed after uranium in the table, but unlike uranium, none of these occurs naturally. All these so-called transuranic elements decay spontaneously by the process of radioactivity, and any that had ever formed along with the rest of the matter in our solar system have now long since decayed. The element directly following uranium in the periodic table was not discovered until 1940, when 'neptunium' was first prepared in the laboratory in tiny proportions by bombarding atoms of uranium with neutrons.

The next element in the table after neptunium is plutonium, named after the planet Pluto, which in turn was named after the Greek god Pluto, ruler of the underworld and brother of Zeus and Poseidon. This planet had been discovered in 1930 by Clyde Tombaugh, who credited the eleven-year-old schoolgirl Venetia Burney with proposing its name. Just ten years later, plutonium was synthesized in the same laboratory that had first synthesized neptunium, at the University of California in Berkeley under the leadership of Glenn T. Seaborg (1912–99). Seaborg relates how he considered the names 'ultimium' and 'extremium' for this new metal, in the mistaken belief that it was the final element in the periodic table. It is fortunate that he decided against these names because, to date, scientists have synthesized twenty-four elements after plutonium. These include berkelium, californium, and americium, as well as element 106, named in honour of Seaborg himself. He became the first living scientist to have an element named after him, and the only person who could address a letter to himself using only the symbols of chemical elements: Sg, Bk, Cf, and Am. On 28 November 2016, Russian physicist Yuri Oganessian became the second living person to have an element named after him—element 118, oganesson, which neatly finishes off the seventh row in the periodic table that starts with element 87, francium.

On 21 October 2003, a new body was detected beyond Neptune by the Palomar Observatory in California and labelled 2003 UB313. On 6 September 2006, the team led by Mike Brown at the California Institute of Technology officially suggested the name Eris, after the goddess of strife and discord, who stirs up jealousy and envy and causes fighting and anger among men (she was famously the cause of the Trojan War). This was a most apt name for the new discovery since it led to the demotion of Pluto as the ninth planet in the solar system. After realizing that there were other bodies orbiting the Sun with similar masses to Pluto (Eris was actually found to be more massive than Pluto was), a new definition of a planet was agreed upon. One of the key requirements was that a planet needed to 'clear its neighbourhood' of other bodies around it, excluding any orbiting moons. In other words, a planet needs to be the dominant gravitational body in that region of space. This is what led to the classification of Pluto, Ceres, and others including the newly discovered Eris as dwarf planets. Pallas failed to meet a further criterion, that it should be nearly round in shape, and so it remains an asteroid. Nevertheless, Pluto, Ceres, and Pallas have all ended up with an element named after them. Eris is now the most massive body in the solar system not (yet) to have its own element.

The Neglected Planet

Before all the controversy over dwarf planets—in fact, before the discovery of Neptune—there remained one true planet in our solar system overlooked, with no element associated with it: Earth. This was perhaps understandable because of the initial geocentric view of the universe. But just before the end of the eighteenth century, this oversight was remedied, once again by Klaproth. He had been sent a very rare sample of a mineral from Transylvania which had proved most troublesome to analyse. It certainly contained gold, but its other main constituent remained elusive. Some thought it might be bismuth; others antimony. In fact, it had proved so much of a problem that the mineral came to be called *aurum paradoxum*, or *metallum problematicum*.

Klaproth writes:

> Among the various products of the mineral kingdom with which Nature has filled the subterraneous parts of Transylvania, those fossils deserve the particular attention of the naturalist which are known by the name of *white gold-ore* and *grey gold-ore*.
>
> Almost all what hitherto has with certainty been known of these minerals, is that they contain *gold* and *silver* in various proportions: but the chemical knowledge of the other constituent parts of them continued involved in doubt and uncertainty.
>
> To fill up this vacancy in the chemical mineralogy, I here submit the experiments I have made with these costly fossils; the principal result of which consists in the discovery of a *new distinct metal*, to which I give the name TELLURIUM, from *tellus*, the latin name of old mother *earth*.[37]

Tellus was the Roman goddess personifying the Earth. Her Greek counterpart was Gaia, who we have already met—the wife of Uranus, who gave the flint scythe to her son Kronos. The elements uranium and tellurium named by Klaproth therefore form a pair named after the god of the sky and the goddess of the Earth.

One characteristic of tellurium that Klaproth notes is that when burning, it is 'particularly distinguished by the somewhat nauseous *radish-like smell* that it emits'. This has since come to be one of the well-documented properties of tellurium; a trial in the nineteenth century in which a volunteer was given half a microgram to ingest (hardly even a visible amount) gave him radishy-garlic breath for thirty hours. More shocking still is the report that a 15 mg dose produced the effect for 237 days.[38]

While Klaproth was the great German mineralogist of his day, Sweden soon produced an equal in the form of Jöns Jacob Berzelius, born some thirty-six years after Klaproth. We've already come across the metal cerium, named after the dwarf planet Ceres; this element was independently discovered by both Klaproth and Berzelius in 1803, but it was the name suggested by Berzelius that survived. (Klaproth had proposed the name ochroite, from the yellow-brown colour of the metal oxide.) Berzelius also invented the universal system of chemical symbols for elements and compounds that we still use today. But it is another element that Berzelius discovered which concerns us now.

On 23 September 1817, Berzelius wrote in a letter how he had discovered traces of Klaproth's tellurium in sulfur residues in a sulfuric acid plant that he had recently purchased. This find was surprising since none of the element had been detected in the minerals used in the manufacturing process. However, in February of the next year, Berzelius wrote correcting himself with even more exciting news: the element was not tellurium, but a new element with properties in between those of sulfur and tellurium. He added: 'When one heats this new substance with a flame, it burns with an azure blue flame, and gives a very strong odor of radishes; it was this odor that made us think it was tellurium...The similarity to tellurium has given me occasion to name the new substance selenium.'[39]

Berzelius's error is understandable; selenium is in the same group in Mendeleev's table as sulfur and tellurium, which means they all have very similar chemical properties. Berzelius tell us, 'I have given the name selenium, derived from Selene (the moon), to recall its analogy with tellurium.' Selene was the Greek goddess of the Moon, and is the counterpart of Luna from Roman mythology.

It is this connection to the Moon that explains the name of the mineral named selenite. Contrary to expectation, selenite is not an ore of selenium; in fact, it does not contain any selenium at all. Selenite is actually a form of gypsum or calcium sulfate, the material used to make plaster casts for broken limbs. Like selenium, it is named after the celestial goddess, but in this case, because it was once thought to wax and wane with the phases of the moon. As Pliny the Elder puts it in the first century AD: 'Selenites is a precious stone, white and transparent, yielding from it a yellow lustre in manner of honey, and representing within it the proportion of the Moone, according as she groweth toward to the full, or decreaseth in the wane against the chaunge.'[40]

The element selenium certainly does not wax and wane with the moon, but it has nonetheless been found to respond to light. Selenium is a semiconductor whose electrical conductivity increases when it is exposed to light. This has been put to use in selenium light meters for photography.

In the next chapter, we turn our attention away from the heavens and look into the very bowels of the Earth, at the sixteenth-century subterranean miners and the strange demons they encountered.

GOBLINS AND DEMONS

There is no danger attending the encreasing number of the metals. Astrological influences are now in no repute among the learned, and we have already more metals than planets within our solar system.

—Cronstedt, 1770[1]

The belief that there were no more than seven metals persisted for hundreds of years, and it was not until the seventeenth century that the inconvenient, inescapable realization came that there were probably many more. I've already mentioned Barba's report from 1640 about the new metal bismuth; it was one of a number of metals or metal-like species that began to be noticed in the sixteenth and seventeenth centuries. In his *History of Metals* from 1671, Webster begins Chapter 27: 'Having now ended our Collections and Discourse of the seven Metals, vulgarly accounted so; we now come to some others, that many do also repute for Metals; and if they be not so, at least they are semi-Metals, and some of them accounted new Metals or Minerals, of that sort that were not known to the Ancients.'[2]

In the chapter Webster speaks of antimony, arsenic, bismuth, cobalt, and zinc. While we now understand these as distinct elements, earlier on there was great confusion, with the names being used for compounds rather than the elements themselves—and, furthermore, the different compounds and elements often being mistaken for each other. This makes unravelling their history all the more complicated. We'll start with Barba's 'Mettal between Tin and Lead, and yet distinct from them both': bismuth.

Bismuth

The first mention of bismuth predates Barba's reference by more than one hundred years. The name appears in its variant spelling, 'wissmad', in what is probably the very first book on mining geology. This was published around the turn of the sixteenth century and attributed to one Ulrich Rülein von Calw, the son of a miller who entered the University of Leipzig in 1485. Ulrich mentions in passing that bismuth ore can be an aid to finding silver, since the latter is often found beneath it. Consequently, miners called bismuth 'the roof of silver'. As Webster later put it in his *History of Metals*, 'The ore from whence it is drawn...is also more black, and of a leaden colour, which sometimes containeth Silver in it, from whence in the places where it is digged up, they gather that Silver is underneath, and the Miners call it the Cooping, or Covering of Silver.'[3]

This helpful property of bismuth was picked up by the sixteenth-century German mining authority Georgius Agricola (1494–1555). He coined the Latin name *bisemutum* and described the metal as *plumbum cinereum* (ash-coloured lead) to distinguish it from tin, white lead (*plumbum candidum*), and true black lead (*plumbum nigrum*). Agricola's first introductory book on mineralogy, *Bermannus*, published in 1530, is in the form of a dialogue between the mineralogist Bermannus and the scholars Nicolaus Ancon and Johannes Neavius as they stroll through the mines.

> Bermannus: Before leaving this place I wish to call your attention to another kind of mineral that belongs to the metals and which, I believe, was unknown to the Ancients. We call it *bisemutum*.
>
> Naevius: Then you believe that there are more than the commonly accepted seven kinds of metals?
>
> Bermannus: I am of the opinion that there are more for this metal our miners call *bisemutum* you cannot correctly call either *plumbum candidum* [tin] or *plumbum nigrum* [lead] since it differs from both and is therefore a third metal.

Its ash colour may have given rise to the name 'bismut', with '*wis mat*' or '*weisse masse*', meaning a 'white mass' or 'white material'. This would be referring not to the element itself, but to ores of bismuth, some of which are white. Bismuth metal has more of a greyish-pink tint, often with a beautiful iridescence owing to a thin layer of the oxide on its surface. Its appearance led to a more romantic proposal for the etymology of bismuth by an associate of Agricola, a Protestant minister named Johannes Mathesius (1504–65) who became interested in mining after reading *Bermannus*. Mathesius would educate his congregation of miners with sermons relating to mining; in one on lead and bismuth, he derives the name of bismuth from '*Wiesen*'—'a meadow'—since the colour of the metal is like the pink blossom of a meadow of clover.

By the mid-sixteenth century, German miners were beginning to prepare metallic bismuth. Since only a low heat is required to melt out any native bismuth from its ore, the process could easily be accomplished on an open fire— as shown in one of the woodcuts accompanying Agricola's most famous book, his *De re metallica*, published in 1556 (Figure 20). Molten bismuth runs out from

Fig. 20. A woodcut from Agricola's *De re metallica* from 1556 showing the production of bismuth.

the fire burning around the ore (G) into iron pans (H and D), giving, on cooling, hemispheres of bismuth (B).

But while metallic bismuth was being prepared in Germany, it was rare to find it outside that country. Even in 1671, Webster states:

> I could never hear of any [bismuth] that was gotten in his Majesties Dominions; and therefore should desire all ingenious Gentlemen that are inquisitive after Minerals, and all other persons that seek or dig for Ores, to inquire if any may be heared of or discovered in this Nation, for it would be a commodity of great worth, for the Metal is very dear. Neither have I ever been so happy, as to be able to procure any of this Ore, and therefore cannot of mine own knowledge give the Reader any satisfaction about the properties of it.[4]

Little did Webster know that metallic bismuth had actually been prepared by German miners in England more than a century earlier. The account books from their operations still exist and describe the sale of copper, lead, and silver prepared in mines in England's Lake District. The entry for 31 January 1569 also includes 14 shillings and sixpence paid to a London carrier for the transportation of four hemispheres of bismuth (perhaps like those depicted by Agricola) weighing 147 pounds, to be sent from Keswick to London and Antwerp.[5] As to how the samples might have been used, a clue comes from another name used in England for bismuth: tin-glass. A late sixteenth-century English text mentions: 'tynne glasse which is that Bisemutum, and that sinder or ashie kinde of leade whereof *Agricola* speaketh'.[6]

The Chemist's Basilisk or Demogorgon

According to Webster, bismuth acquired the name tin-glass because those artists making pewter 'use to mix it with Tin, that it may confer splendor and hardness to it, and that being melted it may run more easily'.[7] The fact that bismuth hardens and gives a shine to tin may explain another name used for the metal, namely the Chimistes basiliske—the basilisk being a mythical serpent capable of turning anyone who looked at it into stone.

The seventeenth-century German chemist Johann Rudolf Glauber (1604–70) discusses the hardening properties of this semi-metal that he calls 'our Demogorgon', named after the dreadful snake-haired sisters from Greek literature whose look turns the beholder to stone. He writes, 'This our Demogorgon hath the virtue even as it comes raw and unprepared out of the Earth to change and meliorate all metals as follows.'[8] He first describes its effects on lead: 'It makes *Saturn* as hard and white as *Lune*, when tinged with it, of which all manner of Vessels and Dishes may be made.' Apparently, the new alloy is almost as good as silver itself: 'it onely wants the sound of *Lune* and enduring of the Test', meaning it would need to be assayed to tell the difference. Its effects on tin ('*Jupiter*') are similar, but with silver, it changes its appearance: 'This Tincture cast upon *Lune*, makes the same Coal-black throughout, so that it is no more like *Lune*…By this means also in times of War, or other danger *Lune* may so be disguised as not to be known for such, and so may be a good way to preserve it from being taken by the Enemy.' It also hardens gold: 'In like manner it makes *Sol* so hard that it can no way be bent or destroyed.' He gives many possible uses for this hardened gold, including for statues, coins, and rings, 'especially such as are designed for the remembrance of Friends, as lasting for ever'. He even suggests, 'Great Princes also might have Armour and Arms made of this hardened *Sol*, which would be much better than any of Iron or Steel, which easily take rust, to which *Sol* is not Subject.'

It was not only in the preparation of pewter that bismuth found metallurgical use. The metal was also used, together with its relative antimony, in making 'type alloy'. This rather specialized application stems from a rare property shared by metallic antimony and bismuth: like water, they are denser in liquid form than in solid form. We are all familiar with the fact that ice floats on water, but broadly speaking, this quality is highly unusual among substances. It is ideal for the preparation of the type blocks used in printing because after the liquid metal is poured into the mould, it expands slightly on solidifying. This ensures a very fine impression in the finished casting.

Unlike its toxic neighbours in the periodic table (polonium, lead, and thallium), compounds of bismuth are not especially poisonous. Despite this, bismuth was not employed in early medicines, possibly because it was not widely known. One of the first popular uses of a bismuth compound was in make-up. The French chemist Moyse Charas describes, in his *Royal Pharmacopœa* of 1678, the preparation of a pure white powder, the 'Magistery of Bismuth' or 'White of Pearl', 'fit for all deformities of the Skin, and to beautify the Complexions of Ladies'.[9] Its use as a cosmetic is described rather disapprovingly by John Henry Pepper in his *Playbook of Metals*, published in the late nineteenth century:

When God's image is defaced with a mineral and poisonous powder, it must of course be for some special purpose; now, the savage tribes chiefly wear paint when they go to battle, to frighten their enemies, and hence the term 'war paint', used by modern writers to express this peculiar decoration worn by a few silly women, is perhaps one of the severest reproofs ever administered to that absurd and foolish practice. The metal bismuth, so called from the German *wiessmate*, or 'blooming meadow', is the one selected for this office; not, of course, in the metallic state, but combined with nitric acid and water, and called the trisnitrate of bismuth, or 'flake white'.[10]

A further anecdote concerning this application is described by John Scoffern in his delightful book from 1839, *Chemistry no Mystery*. During a discussion about the gas hydrogen sulfide (which he calls 'sulfuretted hydrogen'), he relates how it instantly reacts with solutions containing certain metals to give insoluble black precipitates. This reaction may be used to test for the metals, or for the presence of the gas which may sometimes be found in natural spring waters. Scoffern relates:

> It *was* a practice with those ladies who were particularly ambitious of possessing a white skin, to daub themselves with a preparation of the metal bismuth, which is one of these that sulphuretted hydrogen blackens. Now it is represented on creditable authority, that a lady made beautifully white by this preparation, took a bath in the Harrowgate waters, when her fair skin changed in an instant to the most jetty black. You may judge how much was her surprise at this unlooked-for change; uttering a shriek, she is reported to have swooned; and her attendants, on viewing the extraordinary change, almost swooned too, but their fears in some measure subsided on observing that the blackness of the skin could be removed by soap and water. The lady soon recovered from her trance, and derived some consolation from having the true state of things explained to her by her physician, although she was not very well pleased that people should have discovered the philosophy of her white skin.[11]

One final possible etymology for the name bismuth stems from its confusion with the element antimony. Antimony is directly above bismuth in the periodic table, meaning the two elements are in the same family and so have similar chemical properties. The confusion is nicely illustrated by the sixteenth-century physician Philippus Aureolus Theophrastus Bombastus von Hohenheim (ca. 1493–1541), more commonly known as Paracelsus. Paracelsus was rather a colourful character with arguably one of the greatest egos of all time—even his self-appointed name means 'above Celsus', the great scholar from the first century. Whether or not he deserved such accolades, Paracelsus is now credited with popularizing the introduction of minerals—metallic compounds—into medicines for internal use. Paracelsus mentions bismuth, but seems to regard it as a type of antimony:

> There is found a twofold Antimony, one vulgar and black, by which Gold is purified, being molten in it. This hath the nearest affinity with Lead. The other is white, which also is called *Magnesia*, and *Bisemutum*. This hath the greatest affinity with Tin.[12]

It is as a 'second kind of antimony' that bismuth may derive its name. In Arabic antimony is known as *ithmid*, and it has been suggested that the word 'bismuth' simply comes from *bi-ithmid*, 'like antimony'.

Antimony—The Wolf of Metals

Following Paracelsus's popularization of minerals in medicine, the use of antimony attracted more attention and controversy than any other sub-stance. In 1671, Webster noted that 'there is scarcely any one Mineral that is more largely treated of than Antimony'.[13] In 1802 the eminent Scottish chemist Thomas Thomson added, 'No metal, not even mercury nor iron, has attracted so much of the attention of physicians as antimony. One party extolled it as an infallible specific for every disease; while another decried it as a most virulent poison, which ought to be expunged from the list of medicines.'[14]

Antimony is most commonly found in nature as the sulfide mineral stibnite, and occasionally as the element itself; early literature mixes up the two, and it's sometimes hard to tell whether references are being made to the mineral sulfide or to the element. Transcaucasia and Mesopotamia, areas now in the Middle East, were both rich sources of the mineral in ancient times. Crude elemental antim-ony is relatively easy to extract from the sulfide, and a few metallic articles also exist from this same region dating back to the third millennium BC. The appear-ance of the black mineral stibnite is described by Étienne-François Geoffroy in 1736: 'Some is composed of fine shining Lines like Needles, sometimes disposed in regular Ranks, sometimes without any observable Order, which is termed 'Male Antimony'. Some is disposed in thick broad Plates or *Laminæ*, called 'Female Antimony by *Pliny*.'[15] The 'male antimony' is almost certainly the sulfide mineral, but the 'female antimony' might actually refer to the native metal.

Pliny the Elder's description of stibnite in his *Historia Naturalis* from the first century AD is largely taken from the *Materia Medica* (as it is known in Latin) of his Greek contemporary Pedanius Dioscorides. Both authors discuss the main use of stibnite as 'cleansing ye filth & ulcers which are in ye eyes'. In Pliny's words, 'antimony' (probably meaning the mineral) 'has astringent and cooling properties, but its principal use concerns the eyes, and this is why many call it platyophthalmus (eye dilator), since it is the active principle in beauty preparations for dilating the eyes of women. To prevent ulcerations and discharges from the eyes, it is powdered with frankincense and made up with gum.'[16]

This use of stibnite as an eyeliner probably began much earlier in Persia, although it is likely that other black minerals were more commonly used. Analyses of ancient Egyptian samples of eyeliner (called *mestem*) have revealed that they were often galena (lead sulfide) or even soot. However, it is the Egyptian name *mestem* or *stem* which gave rise to the Greek names *stimmi* and *stibi* and the Latin *stibium*, which was used to refer to the mineral. This Latin root '*stibium*' led Berzelius in the early nineteenth century to use the symbol Sb for antimony, and it is still used by modern chemists.

It was not only the sulfides of antimony and lead that were confused with one another—the elements themselves have also been mixed up. Pliny, for example, describes how to make various preparations from stibnite by roasting it in air (which would convert it to the oxide) and then heating with, among other things, 'lumps of cow-dung'. He warns, 'it is a matter of prime importance that the roasting should not be too vigorous, lest the product turn to lead'. This 'lead' is undoubtedly actually metallic antimony, which would be produced by the chemical reduction of the antimony oxide with the burnt organic matter.

Both lead and antimony were used in early metallurgy to purify gold. In a process known as cupellation, the impure gold would be roasted with lead in a porous vessel known as a cupel. Remarkably, this process, relatively unchanged over the centuries, is still carried out today during the assaying of gold. Perhaps through analogy with Saturn consuming his children, the lead is said to devour all the metallic impurities. In the high temperatures of the assay furnace the lead and impurities form molten oxides which are absorbed into the cupel itself, leaving the gold behind. If any silver is present, this is also left unchanged, mixed with the gold. Carrying out the process with stibnite instead of lead also removes any silver contained in the sample. This explains some of the names given to antimony by early alchemists and metallurgists, as explained by Boerhaave in 1727:

> The chemists have two kinds of lead, or *Saturn*: *viz.* the *Saturn* of Diana, or common lead; and that of *Sol*, called also the *Saturn of the Philosophers*, which is

antimony. None but gold and silver resist the first; and none but gold alone the second; each they term *Lavacrum Leprosorum*, or the *Lepers Bath*; from which they arise the cleaner: intimating hereby, that all other metals by them deem'd leprous, ☉ and ☾ excepted, fused in the same cupel with lead, or antimony, fly off in fume…

As to antimony, a quantity thereof being put in a cupel, along with pure gold, and the whole fused, and kept in a strong fire; the antimony all evaporates, and leaves the gold alone: which does not hold of any other metal, not even *silver* itself. Whence antimony is particularly called *Balneum Solis*, the sun's bath; and *Lavamen solis Regis*; *Devorator*; and *Lupus Metallorum*, &c.[17]

Lupus Metallorum translates as 'the Wolf of Metals'. In the English translation of his pharmacopeia from 1669, Schroeder says of antimony, 'It hath divers Names, and is called a Wolf, because it devours all Metals but Gold; and Proteus, because it takes all colours by fire; or the Root or Mineral of Metals; and Saturn Philosophical, because like Lead; and because they dream that the Philosophers Stone is to be made by it.'[18]

This process of purification by cupellation is depicted in an image based on a work by Basil Valentine (Figure 21). The King and Queen represent the regal

Fig. 21. The process of cupellation as depicted in *The Twelve Keys of Basil Valentine*. This engraving is taken from the *Viridarium Chymicum* of Daniel Stolcius, 1624.

metals gold and silver, which can be purified using antimony (represented by the wolf) or lead (represented by Saturn, identifiable with his scythe). The process is described by Valentine: '[T]ake the most ravenous grey Wolf, which by reason of his Name is subject to valorous *Mars*, but by the Genesis of his Nativity he is the Son of old *Saturn*, found in Mountains & in Vallies of the World: He is very hungry, cast unto him the Kings body, that he may be nourished by it; and when he hath devoured the King, make a great fire, into which cast the Wolf, that he be quite burned, then will the King be at liberty again.'[19] The process of restoring the King—i.e. recovering the purified gold—is also shown in Figure 22.

Basil Valentine is one of the most curious characters connected with antimony's history. He is said to have been born in Germany in 1394 and to have entered the St Peter's Benedictine monastery at Erfurt by 1413. Unfortunately, there seems to be no concrete evidence for his existence, and he is thought to

Fig. 22. The purification of gold as shown in an engraving in Michael Maier's *Atalanta Fugiens*, first published in 1617.

be entirely fictional. None of his writings exist before around 1600, but this is conveniently explained, according to the title page of the *Last Will and Testament*, published in 1670, by the fact that he deliberately concealed his *magnum opus* 'under a Table of Marble, behind the High-Altar of the Cathedral Church, in the Imperial City of *Erford*: leaving it there to be found by him, whom Gods Providence should make worthy of it'. Legend has it that the texts were revealed after lightning struck the altar. It is not only the late sudden appearance of his works which is highly suspicious; the content is also far ahead of the early fifteenth century, including a reference to America.

Basil Valentine's Triumphant Chariot

Whether or not Valentine actually existed, in 1604, what has been described as one of the earliest monographs on a chemical element was published under his name: *Triumph Wagen Antimonii*, first translated into English in 1660 as *The Triumphant Chariot of Antimony*. Its rather pompous title is explained by Valentine:

> How ill it [antimony] hath been spoken of in this our Time, is well known to many; and of how serviceable use it hath been, in the Cure of very many deplorable Diseases; within twenty years last past, is as well understood by most of the industriously laborious Physicians of this City; who can, and dayly do, whensoever they meet with Men of like Industry, testify for the Author, that unto ANTIMONY is not undeservedly assigned a CHARIOT TRIUMPHANT. For none were permitted to enter *Rome* in a Triumphant Chariot, that had not slayn at least five thousand *Enemies*, and obtained an intire Victory.[20]

One curious remark from the book has given rise to a charming legend associated with Valentine and the origin of the name antimony. He first describes the mode of action of antimony:

> Moreover be it known to all, that Antimony doth not onely purge Gold and separate all extraneous additions therefrom, but performs the same operation in the bodyes of men, and other living creatures, which I shall prove by an homely example. If a householder intends to fatten a beast, but especially an hog, let him give him in his meat (three dayes before he shuts him up) halfe a dragme of crude Antimony, by which means his appetite to his meat will be, whetted, and stird up within him, and heel soone grow fat; and if he hath any hurtful quality or disease in his liver, or be leaprous, he shall be healed.[21]

Valentine qualifies his story as being more of a parable intended for the less intellectually gifted, and ends with a warning:

> This example will seem somewhat grosse to the ears of delicate men; but I intended it for illiterate men, or country people, in whose brains the more subtile Philosophy is a meer stranger, that they may discerne that experimentally, which for examples sake I have made use of, that so they may the sooner credit my other writings, wherein I speak more abstrusely: But because theres a great difference between the bodies of men and beasts, I have no intent (by this example here induced) that crude Antimony should be given to men also; because that the beasts are able to bear and concoct much crude meats; which is not permitted to the tender nature and complexion of man to doe.

How this story is connected to the origin of the name antimony, or *antimoine* in French, is related by the seventeenth-century French apothecary Pierre Pomet:

> It acquir'd the Name of *Antimony*, according to the Opinion of some, from a *German* Monk, the aforesaid *Valentine*, who, in his Search after the Philosophers Stone, was wont to make much Use of it for the more ready fluxing his Metals; and throwing a Parcel of it to some Swine, he observ'd that they had eaten it, and were thereby purg'd very violently, but afterwards grew the fatter upon it; which made him harbour an Opinion, that the same sort of Carthartick, exhibited to those of his own Fraternity, might do them much Service; but his Experiment succeeded so ill, that every one who took of it died. This therefore was the reason of this Mineral being call'd *Antimony*, as being destructive of the Monks.[22]

It was soon pointed out that while this theory may hold in France, where the word for monk is '*moine*', a fifteenth-century Basil Valentine would have used the term then in vogue for antimony in his native Germany: '*spiessglas*', a name derived from the fact that antimony oxide could be melted at high temperatures and poured out onto marble to form, on cooling, a clear glass-like substance. More concrete evidence that the story is apocryphal comes from the fact that the term 'antimony' had been used hundreds of years before Valentine's supposed birth, as we shall see.

There have been a few other spurious etymologies. A variation of the 'anti-monk' theory is that the name comes from 'anti-monos', reflecting the fact that the mineral is usually found mixed in with other minerals. As Geoffroy puts it in 1736, 'Antimony is sometimes found in a particular Oar, but most commonly mixed with other Metals; and hence its Name may have been derived. Antimony being the same as ἀντίμονον [antimonon], an Enemy to Solitude.'[23] Still more fanciful suggestions include a derivation from the Greek *anthemonion*, meaning 'to come to flower', from the plant-like crystalline structure of stibnite; 'anti-minium', reflecting its use as an alternative eyeliner to the orange-red lead mineral pigment *minium*; from *antimonetus*, arising from the idea that it would be hard to counterfeit money using alloys containing antimony, because the metal would be too brittle. Curiously, this did not put off the Chinese in the province of Kweichow from trying in the 1930s: they used local minerals to produce an alloy of lead and antimony, but the coins were unpopular and never caught on.

Perhaps the earliest use of the word 'antimony' as distinct from *stibium* surfaces in the eleventh century, when Constantine the African was translating Arabic medical texts into Latin. A twelfth-century manuscript copy of his *Liber De Gradibus* (Book of Degrees) preserved in the Wren library of Trinity College Cambridge uses the term 'antimonium' before discussing the medicinal properties of what was undoubtedly stibnite, antimony sulfide. Texts such as these soon morphed into the early pharmacopoeias and herbals that date from the beginning of printing. One of the earliest printed in English is *The Grete Herball*, published in 1526. This is a remarkable book, following the same style as the earlier Latin *Hortis Sanitatis* from the end of the fifteenth century. The *Grete Herball* gives descriptions and medicinal uses of many plants and a number of minerals. Each entry begins with a crude woodcut to illustrate the topic, although many are used more than once for different entries. It is hard to imagine what the picture for antimony actually represents—it looks more like coffee beans than metal—but it is interesting that this same woodcut is used for the entry on lead.

The text goes on to describe the medicinal uses of antimony, all of which are external: against 'canker' and 'polypes', since the powder 'wasteth the deed flesshe'; for 'the spot in the eye', 'agaynst bledynge of ye nose', and 'for emorroides'. But it is for *internal* use that antimony was to gain notoriety in the seventeenth century.

As it is rather poisonous, the body unsurprisingly tries to expel any ingested compounds of antimony. While there was one early school of thought that compounds of antimony were therefore simply best avoided, another reasoned that good use could be made of them as emetics. This idea is depicted rather violently in the woodcut from Barlet (Figure 23), and described by Glauber in 1651:

ANTIMOINE.

Fig. 23. A woodcut from 1673 graphically depicting the use of antimony as a purgative.

> There is no one that can deny that Antimony is the most excellent of all vomitives…Antimony reduced into glass [antimony oxide] is sufficient to purge the stomach and bowels from all corrupt humors, and that without all danger (being rightly administred) as well by vomit as by stoole, by which means many grievous imminent diseases are not only prevented, but also presently are cured…After which manner being given it attracts from all the bowels all vitious humors, and evacuates them aswel upward as downward without danger…[24]

This powerful emetic could simply be formed by adding some powdered 'glass of antimony' (fused antimony oxide) to wine, allowing the solid to settle and decanting off the 'stibiated wine' or *vinum emeticum*.[25] It was even possible to reuse the solid in further preparations. A most inventive twist on this procedure was to make the drinking cup itself from the vomit-inducing compound. There were two methods of accomplishing this: one was to produce an antimony glaze over an earthen vessel, the second was to make the whole cup from metallic antimony. Glauber tells us:

> Let him that useth the aforesaid cups infuse one or two ounces of wine, and set them a whole night in some warm place, and the wine will attract from the

glass so much as doth suffice it, which afterwards being drunk in a morning, doth perform the same as infusion made with the powder of Stibium.[26]

He adds that 'one cup is sufficient for the master of a family, with his whole family for all the dayes of their life'. The vessels were evidently highly valued and passed on as family heirlooms. One such cup associated with the eight-eenth-century explorer Captain James Cook was sold at auction in 2005 for an incredible £220,800.

A cheaper but equally long-lasting alternative was the so-called *perpetual pill*. These were made from the molten metal formed 'into Bullets of the bigness of a Pill'.[27] Its use was simple: 'When one swallows the *Perpetual Pill*, it passes by its own weight, and purges downwards; it is washed and given as before, and so on continually.'[28]

Glauber also describes how emetic cups may be prepared without the use of antimony. Getting rid of this poisonous element sounds good, until you find out that it was replaced with an even more toxic element: arsenic. As arsenic is situated above antimony in the periodic table, they share many proper-ties. This also means that traces of arsenic are often difficult to remove from antimony compounds, and this route has been suggested for the discovery of residues of arsenic in the hair of King George III of England. The King had a well-documented bout of mental illness, thought to be due to the hereditary disease porphyria. The onset of the disease could in turn have been triggered by residues of arsenic present in the antimony medicines administered to him.

Arsenic

Arsenic has been known for thousands of years, but again, in the form of its compounds rather than the free element. The name comes from 'ἀρσενικόν (arsenikon)', the word the Greeks used for the bright yellow mineral *orpiment*. Confusion with the Greek word for 'male', 'ἀρρενικός (arrenikos)' or 'ἀρσενικός (arsenikos)', has led to a number of spurious etymologies trying to ascribe potent, masculine properties to the element. It is more probable that the Greek term actually derives from the Arabic term for the same mineral, '*al-zarnikh, zarnik*' meaning 'gold-coloured'. The 1634 translation of Pliny's *Natural History* from the first century AD tells us that orpiment is 'a minerall digged out of the ground in Syria, where it lyeth verie ebb, and the painters use it much: in colour it resembleth gold, but brittle it is in substance like as glasse-stones'.[29]

This use as a golden pigment explains the current name of the mineral, orpiment, derived from the Latin '*auri-pigmentum*'. Orpiment is a sulfide mineral
of arsenic and has the chemical formula As_2S_3. Both Dioscorides and Pliny
mention a second, bright red form of arsenic they call 'sandarach', which we
identify with the mineral realgar. This too is a sulfide of arsenic, but with the
formula As_4S_4. The name realgar also comes from Arabic; '*rahj al-gar*', literally
meaning 'dust of the cave'. This may have been confused with '*rahj al-fa*', meaning 'rat powder', although the term '*samm al-far*' ('rat poison') was more commonly used when the mineral was used as a rodenticide.

Despite being poisons, both minerals were used as medicines. Pliny and
Dioscorides both describe arsenic's external use to eat away protrusions of
the flesh, but also, more worryingly, state that 'taken in honey, it cleanses the
throat, and renders the voice clear and tuneful'.[30]

Far more poisonous than either of the sulfides is the deadly 'white arsenic',
arsenic oxide, formed on burning the sulfide minerals in air. Lemery mentions
all three forms of arsenic encountered in the seventeenth century, but cautions
against its use:

> *Arsenick* is a Mineral Body consisting of much Sulphur, and some Caustick salts.
> There are three sorts of it, White that keeps the name of *Arsenick*, Yellow called
> *Auripigmentum*, or Yellow *Orpin*, and Red called *Realgal*, or *Sandaracha*, the White
> is the strongest of all.
>
> None of these *Arsenicks* can be given inwardly, though several persons that
> have ventur'd to use the White, do pretend to have cured with it divers Diseases,
> and among others the Quartan-Ague. They venture to give of it as far as four
> grains, in a great deal of Water, and after that manner it will make one Vomit,
> like *Antimony*. But I can by no means allow of this *Febrifugum*, and would never
> advise any body to use so dangerous a Remedy. Nature doth furnish us with
> Medicins enough of Conscience to provoke Vomiting without recourse to
> *Arsenick*. It is used outwardly with sufficient success, because it eats up
> proud flesh.[31]

The other two members of this family of elements are nitrogen and phosphorus.
Arsenic works as a poison because it most closely resembles phosphorus, an
element crucial to life. The structure of DNA includes a phosphate backbone,
and phosphate also appears in ATP (adenosine triphosphate), the energy cur-
rency used throughout the body. Arsenic can mimic the phosphorus in phosphates, but its subtly different properties eventually sabotage the finely tuned
biological systems, ultimately causing death.

Curiously enough, it is the mimicking property of another arsenic-containing, flesh-eating mineral that gave us the name of the next element we are concerned with: cobalt.

Gnomes and Goblins

In his *Treatise of the Fossil, Vegetable, and Animal Substances, That are Made Use of in
Physick* from 1736, after discussing the minerals orpiment and realgar, Geoffroy
describes a different kind of arsenic:

> Arsenick properly so called, is a Substance extracted from an Oar found in
> *Saxony* and *Bohemia*, named *Cobalt…German* Cobalt of the Shops, *Cadmia
> Metallica* of *Agricola*, is a ponderous, hard, fossil Substance, almost black, not
> unlike Antimony or some Kinds of Pyrites, emitting a strong sulphureous
> Smell when burnt. It is dug out of Mines in *Saxony*, near *Goslar*; in *Bohemia*, in
> the Valley of *Joachim*; and in *England* in the *Mendip* Hills, in great Quantities. It
> has so strong a Corrosive Quality as sometimes to burn and ulcerate the Hands
> and Feet of the Miners, and is a deadly Poison for all known Animals.[32]

The name 'cobalt' may derive from the word '*cobathia*', which was what the
ancient Greeks called the poisonous smoke of white arsenic oxide formed
when arsenic ores are roasted in air. Geoffroy describes this process of roasting
arsenic ores, but of importance here is not the collected white smoke of arsenic
oxide, but what is left behind once the arsenic is driven off. This black residue,
which we now know as cobalt oxide, was pulverized and mixed with powdered
flint stones moistened with a little water, eventually forming a solid mass called
zaffera. This substance was highly prized for making blue glass and pottery, a
purpose for which it is still used today.

Cobalt compounds have been used in making blue glass and enamels since
the second millennium BC. In ancient Egyptian times, the cobalt was obtained
from the Kharga Oasis in the Libyan Desert in central Egypt, where it occurs
in small amounts in alum deposits (crude potassium aluminium sulfate).
Analyses of different samples of blue pottery and glasses show that in the early
sixteenth century a new source was being used for the cobalt pigment—one
containing arsenic. This was what the German miners came across, minerals
such as cobaltite consisting of cobalt, arsenic, and sulfur. Initially the miners
did not know what to make of this new ore since they could not smelt it to

extract metal. It seemed in some way bewitched, and this is said to have given rise to its name, as John Beckmann relates in his *History of Inventions and Discoveries* from 1797:

About the end of the 15th century, cobalt [mineral] appears to have been dug up in great quantity in the mines on the borders of Saxony and Bohemia, discovered not long before that period. As it was not known at first to what use it could be applied, it was thrown aside as a useless mineral. The miners had an aversion to it, not only because it gave them much fruitless labour, but because it often proved prejudicial to their health by the arsenical particles with which it was combined; and it appears even that the mineralogical name *cobalt* then first took its rise. At any rate, I have never met with it before the beginning of the sixteenth century; and Mathesius and Agricola seem to have first used it in their writings. Frisch derives it from the Bohemian word '*kow*', which signifies metal; but the conjecture that it was formed from *cobalus*, which was the name of a spirit that, according to the superstitious notions of the times, haunted mines, destroyed the labours of the miners, and often gave them a great deal of trouble, is more probable; and there is reason to think that the latter is borrowed from the Greek. The miners, perhaps gave this name to the mineral out of a joke, because it thwarted them as much as the supposed spirit, by exciting false hopes and rendering their labour often fruitless. It was once customary, therefore to introduce into the church service a prayer that God would preserve miners and their works from *kobolts* and spirits.[33]

The Mathesius referred to is the minister we encountered earlier in our discussions on bismuth. He refers to cobalt ore in his tenth sermon: 'Ye miners call it *kobolt*; the Germans call the black devil and the old devil's whores and hags, old and black *kobel*, which by their witchcraft do injury to people and to their cattle.'[34] As Beckmann states, the word for the demon, kobold, may derive from the Greek '*kobalos*', a mischievous satyr fond of imitating people.

In his book *Bermannus* from 1530, Agricola mentions a mineral he calls 'cobaltum', saying that 'very often it possesses an extraordinary corrosive quality so that it will eat into the hands and feet of workmen unless they take careful precautions against it'.[35] In 1549 Agricola published a book on subterranean animals—*De Animantibus Subterraneis*—and at the end of this, he mentions the demons found in mines. Some are cruel and terrible to behold, and these annoy and hurt the miners. He mentions one called Annebergius, 'who only with his breath killed more than twelve labourers in a cave called Corona Rosacea'. Apparently, the demon appeared as a horse with the poisonous gas issuing from his mouth. Agricola then gives a very full description of the *cobalos*, who, he informs us, are not particularly evil:

> Then there are the gentle kind which the Germans as well as the Greeks call *cobalos*, because they mimic men. They appear to laugh with glee and pretend to do much, but really do nothing. They are called little miners, because of their dwarfish stature, which is about two feet. They are venerable looking and are clothed like miners in a filleted garment with a leather apron about their loins. This kind does often not trouble the miners, but they idle about in the shafts and tunnels and really do nothing, although they pretend to be busy in all kinds of labour, sometimes digging ore, and sometimes putting into buckets that which has been dug. Sometimes they throw pebbles at the workmen, but they rarely injure them unless the workmen first ridicule or curse them. They are not very dissimilar to Goblins, which occasionally appear to men when they go to or from their day's work, or when they attend their cattle.[36]

It's hard to imagine what it must have been like to be a miner more than a few hundred years ago. There was no reliable lighting or ventilation; the mines could collapse at any point and crush the miners; they could be poisoned by invisible vapours or blown up by the ignition of pockets of flammable gas. Add to this the stifling heat and the fact that some of the minerals themselves were poisonous and corrosive, and it really must have seemed to the miners that they were venturing into hell. It is hardly surprising, therefore, that they reported encountering demons in these subterranean pits. In 1635, English playwright and author Thomas Heywood published, in verse, *The hierarchie of the blessed angells*, which also includes a description of these mining demons and their activities:

> Subterren Spirits they are therefore styl'd,
> Because that bee'ng th' upper earth exyl'd,
> Their habitations and aboads they keepe
> In Con-caves, Pits, Vaults, Dens, and Cavernes deepe;
> And these *Trithemius* doth hold argument
> To be of all the rest most pestilent:
> And that such Daemons commonly invade
> Those chiefely that in Mines and Mettals trade;
> Either by sudden putting out their lamps,
> Or else by raising suffocating damps,
> Whose deadly vapors stifle lab'ring men:
> And such were oft knowne in *Trophonius* den.

Likewise in *Nicaragua*, a rich Myne
In the West-Indies; for which it hath ly'ne
Long time forsaken. Great *Olaus* writes,
The parts Septentrionall are with these Sp'ryts
Much haunted, where are seen an infinit store
About the places where they dig for Oare.
The Greeks and Germans call them *Cobali*.
Others (because not full three hand-fulls hye)
Nick-name them Mountaine-Dwarfes; who often stand
Officious by the Treasure-delvers hand,
Seeming most busie, infinit paines to take,
And in the hard rocks deepe incision make,
To search the mettals veines, the ropes to fit,
Turne round the wheeles, and nothing pretermit
To helpe their labour; up or downe to winde
The full or empty basket: when they finde
The least Oare scatter'd, then they skip and leape,
To gather't thriftily into one heape.
Yet of that worke though they have seeming care,
They in effect bring all things out of square,
They breake the ladders, and the cords untwist,
Stealing the workmens tooles, and where they list
Hide them, with mighty stones the pits mouth stop,
And (as below the earth they underprop)
The Timber to remove they force and strive,
With full intent to bury them alive;
Raise stinking fogs, and with pretence to further
The poore mens taske, aime at their wracke and murther.[37]

One of the sources used by Heywood for his description is *Historia de Gentibus Septentrionalibus* (Description of the Northern Peoples) by Olaus Magnus, published in 1555. This book describes the customs of the Nordic people and was immensely popular throughout Europe, in part because of the fine woodcuts that illustrate the different entries. The one accompanying his description of the mining demons (Figure 24) shows the miner hard at work on the left of the mine shaft and the demon, pitch-black, on the right.

Perhaps to raise their spirits as they descended into hell, the miners would occasionally sing—in his book of sermons, Mathesius includes a number of mining hymns specifically for this purpose. It may have been images of such singing miners and the diminutive sprites they encountered that provided the inspiration for the tale of Snow White and the Seven Dwarfs; the authors, the

Fig. 24. A sixteenth-century mining demon as depicted in the *Historia de Gentibus Septentrionalibus Historia de Gentibus Septentrionalibus* of Olaus Magus.

nineteenth-century Grimm brothers, were noted linguists and philologists who collected such tales of folklore and were well acquainted with the 'Kobold demons'.

The Grimm brothers were also responsible for popularizing another demon-related etymology for a chemical element: nickel. According to the myth, early German miners came across another arsenic-containing ore that puzzled them since it resembled ore of copper and yet they were unable to extract any metal from it. The miners asserted that a 'nickel', a type of goblin, had robbed the ore of its metal, so they gave the mineral the name *kupfernickel*—'the devil's copper'. The use of the term 'nickel' for a goblin could be related to a diminutive form of 'Old Nick'—a reference to the devil himself.

Far from being bewitched, the mineral yielded no copper simply because it did not contain any. Exactly what it did contain was not understood until 1754, when the Swedish mineralogist Axel Fredrik Cronstedt (1722–65) isolated a new metal from the mineral and gave it the truncated name nickel. The story is described by fellow Swede Torbern Bergman in 1784:

There is found in the parts of Germany which abound in metals, an ore which is called *kupfer-nickel*, sometimes grey, but often of a reddish yellow colour, and polished. This name it probably first got and still retains from this circumstance, that although it has the appearance of containing copper, yet not the smallest particle of that metal can be extracted from it, even by fire. The first account we had of it was from V. Hierne, in a book published in 1694, written in the Swedish language, concerning the discovery of ores and other mineral substances...Mr Cronstedt first undertook an accurate examination of this mineral, and by many experiments, which were published in the years 1751 and 1754, shewed that it contained a new semimetal, to which he gave the name of *nickel*.[38]

Other suggestions have been made to explain how *kupfernickel* got its name. While the '*kupfer*' part is clearly a reference to the ore's resemblance to copper minerals, another suggestion for the *nickel* part is that it is an abbreviation of arsenic, which the mineral also contains. More probable is that it derives from the Latin term '*nichilus*', which was used for the minerals agate and occasionally onyx. This in turn might have come from a much earlier word, '*knock*', used to describe a mountain or hill, such as the summit Nockstein in Salzburg. The word 'nickel' might have been used to signify a small fragment of rock, and a similar corruption perhaps gave us the word 'nugget'.

There was one further twist concerning the names of cobalt and nickel. In 1889 two German chemists, Gerhard Krüss and G. W. Schmidt, proposed that cobalt and nickel were not pure metals, but contained approximately 2–3 per cent of a new element. This idea came from the fact that their atomic masses, mischievously, were not in the order demanded by Mendeleev's new Periodic System. Mendeleev arranged his elements according to increasing atomic mass and looked for the repeating patterns that emerged. However, he realized that a couple of pairings (tellurium/iodine and cobalt/nickel) needed the elements to be arranged with the heavier element before the lighter one in order for them to be placed in the correct groups. This misdirected chemists to suspect that the atomic masses were wrong. (The real reason, discovered decades later, was that the elements were actually ordered by the increasing number of protons that the atoms contained—their so-called atomic number.) Krüss and Schmidt even gave a name to their supposedly new metal responsible for inflating the mass of cobalt. Wanting to reflect its similarities to the impish elements it accompanied, they named it 'gnomium'. Sadly, it later turned out to be no more than a fictional spirit.

Nickel and cobalt were not the only species fooling the miners and chemists; according to the Grimm brothers, the mineral known as blende derived its very name from the German word '*blenden*', 'to deceive'. In the nineteenth century, this mineral was given the name sphalerite, from the Greek for 'deceitful'. This 'deceiving ore' brings us to our next metal, zinc.

Zinc

Richard Watson, the eighteenth-century Regius Professor of Chemistry at the University of Cambridge, writes that blende was given its name by the German miners 'from its blinding, or misleading appearance; it looking like an ore of lead, but yielding (as was formerly thought) no metallic substance of any kind'.[39] The lead mineral this zinc ore resembles was known by Pliny as *galena* (lead sulfide), from a Greek word signifying 'to shine'. 'Zinc blende' was also known as pseudo-galena or mock lead; English miners knew it as blackjack. Watson relates: 'Black jack resembles lead ore so much, that the miners sometimes succeed in selling, to inexperienced smelters, black jack instead of lead ore; I have heard of the fraud being carried to so great an extent in Derbyshire, that from a ton of ore there was not obtained above a few ounces of lead; though a ton of unadulterated lead ore yields in Derbyshire, at an average, 14 or 15 hundred weight of lead.'[40]

We now know the mineral blende is composed of zinc and sulfur, but this was not realized until many centuries after blende had first been used, alongside another zinc ore known as cadmea. This latter ore—cadmea, cadmia, or Cadmean earth—was named after the citadel of ancient Thebes in Greece, founded, according to legend, by Cadmus, the Greek hero who gave us the alphabet. Unfortunately the name cadmia and a possible corruption of it, calamine, have been used variously for zinc carbonate, zinc silicate, and zinc oxide; in the seventeenth century, 'cadmia' could also refer to cobalt and arsenic minerals. To add to the confusion, the mineral now has an element named after it: cadmium, discovered as an impurity in the zinc ore in 1817 by the German chemist Friedrich Stromeyer.

Cadmia, and to a lesser extent blende, were used in the production of the highly prized golden alloy, brass—a mixture of copper and zinc. Significant quantities of brass were produced in the first century AD, when the metal was used in making Roman coins. Brass was formed by heating copper metal with cadmia and charcoal. Zinc is momentarily formed in this process, but instantly dissolves in the copper to form the brass.

Brass can also be made simply by melting the two metals together, but this was not possible until the late seventeenth and early eighteenth centuries, when zinc metal became more widely available—thousands of years after the original seven metals had been known. Prior to this period, only rare samples of the metal were occasionally encountered. This is not because zinc itself is a rare element—it is more abundant in the earth's crust than all of the original seven except iron. The reason lies in how the metals are obtained. While it is possible to find some of the less reactive metals such as gold, silver, and sometimes even copper 'native'—i.e. as the free element—the more reactive elements (such as iron, tin, and zinc) are usually found as minerals combined with other elements such as oxygen or sulfur. In order to isolate the pure metal, it needs to be freed from whatever elements it is combined with; this is the process of smelting. For instance, 'roasting' the green copper ore malachite in air causes it to decompose to black copper oxide. On heating this with charcoal (impure carbon), the carbon steals away the oxygen, forming the gases carbon monoxide and carbon dioxide, and leaving behind the metallic copper. The carbon and carbon monoxide gas are said to reduce the metal oxide to the metal. The same idea applies to the production of iron, except that much higher temperatures are needed for this reaction to work (which is why the more technologically advanced Iron Age necessarily followed the easier-to-achieve Copper Age or Bronze Age).

The reason why no zinc was easily produced was because the temperatures in the furnace—typically in excess of 1200° C—were high enough to cause any zinc that was ever formed to boil away, its boiling point being around 900° C, much lower than the other metals. What is more, once formed, the zinc would quickly react with the oxygen in the air to form a fine white powder of zinc oxide, which often collected in the chimney of the furnace or fell like an ash to the floor. This impure zinc oxide was variously known to the first-century authors Dioscorides and Pliny as *tutia* or *tutty* (derived from the Persian name for this substance); *pompholyx* (from the Greek meaning a blister, in reference to its appearance as it forms in the chimney); or *spodos* (from the Greek for dust or ash). In *The Moste Excellent Workes of Chirurgerye*, published in English in 1543, the Italian physician Giovanni da Vigo writes in the final section, 'The Interpretacion of the straunge wordes', that '*Tutia* is called in Greke, Pompholix, that is to saye, a bubble. For it is that, that bubbleth up in brasse, whan it is boyled, and cleveth to the sydes, or cover of the fornace.'[41]

The formation of pompholyx and spodium are described by Dioscorides as follows (taken from the 1655 English translation by John Goodyer, *The Greek Herbal of Dioscorides*):

> Pompholyx doth differ from Spodium specifically, for it hath not a genericall difference, for ye Spodos is somewhat black & for ye most part heavier, being full of motes and hairs and earth, being as it were ye scraping & shaving of ye

floors & hearths in the brass-finers' shops. But Pompholix is fatt & white, & withall most light, so that it can fly into the air. Of this there are two kinds, one of ye colour of ye air, & somewhat fat, but the other very white, and having ye heighth of lightness. But ye white becomes Pompholix when in ye working and finishing of ye brass, ye brass-finers do sprinkle on ye bruised Cadmia the thicker, willing to have it ye better; for ye smoke that is carried up from this, being most white, is turn'd into Pompholyx. But Pompholyx is not made only from ye working and matter of brass, but also from Cadmia purposely blown with ye bellows for ye making of it. And it is made thus … the coals are put into ye furnace and are kindled, afterward ye workman standing by doth sprinkle on from the places over the head of ye furnace the Cadmia being beaten small. And ye servant that is under withall doth do ye same, & casteth on more coals, until all ye Cadmia that was laid on be consumed, so that by the burning, the thin & light part is carried into the upper room, & sticks to the walls & to ye roof thereof. But the body that is made of those things carried up, at ye beginning indeed is like to ye bubbles standing upon ye waters, but at last more increase coming, it is like to fleeces of wool. But that which is heavier goeth into ye places under foot & is dispersed about, some to ye furnace, & some to ye floor of ye house. This is thought to be worse than that of thin parts, because it is earthy & full of filth by ye gathering of it. And some think that ye aforesaid Spodos is made only thus.

Agricola described this type of artificial cadmia that forms in the furnaces as *cadmia fornacum*. He also states that, 'since the *cadmia* that forms on the iron rods in a furnace is in hollow masses, it first took its name from the hollow reed *calamus*'.[42] This is the calamine referred to earlier. The process of collecting the tutty is illustrated in a German mining book by Lazarus Ercker, published at the end of the sixteenth century (Figure 25).

If there is little or no oxygen present for the zinc vapour to react with, it is also possible for some zinc metal to collect in the chimney. Small samples of the metal were probably formed in this way, but the first large-scale production of zinc metal was carried out in India in settlements around Zawar in Rajasthan, from about the thirteenth century. The technology was taken to China, where production of the metal began from the sixteenth century. Samples of this new metal began to migrate to the West soon after, and one suggestion for the origin of the name zinc comes from the Arabic *sini* and the Persian *cini*, used to refer to metals from China. Presumably because of its scarcity, early examples seem to have been regarded as priceless treasures. For example, amid the massive emeralds, diamonds and the 48 kg solid gold candlesticks on display in the treasury of the Topkapi Palace in Istanbul are a number of zinc drinking goblets,

Fig. 25. Woodcut from Ercker (1580), showing the collection of cadmia or tutty by scraping the chimney of a furnace.

known as *tutiya*, dating from the mid-sixteenth century. The metal from which these were fashioned almost certainly came from India.

Zinc was first imported into Europe in the late sixteenth and early seventeenth centuries under the name *tutenag*, or sometimes with the Dutch name *spiauter*, which became *spelter* in English. Understanding of the metal's nature was limited at this time. In the *Waka Sanzai Zue* (*Chinese and Japaese Universal Encyclopaedia*) from 1712, which in turn was based on an earlier work from a hundred years earlier, the authors state: 'We really do not quite know what this (metal) is, but it belongs to the category of lead, wherefore it is called "inferior lead" (*ya chhien*).'[43] They also tell us that it is sometimes called *totamu* or *tutenag*, 'a word derived from some foreign language'. This name probably came into

China from the south Indian word 'tutthanaga', used for certain ores including zinc carbonate, and related to the tutty we encountered previously. Being unfamiliar to readers of English, the word 'tutenag' often mutated; in The Present State of Great Britain, published in 1707, it is reported that in addition to silk, musk, and chinaware, the Scots also import from China the metals copper, gold, quicksilver, and 'Tooth and Egg'.

In 1673, Robert Boyle (1627–91), working in Oxford, described the effects of heating a number of different metals in air. He writes: 'Among our various tryals upon common Metals, we thought fit to make one or two upon a Metal brought us from the East-Indies, and there call'd Tutenâg, which name being unknown to our European Chymists, I have elsewhere endeavoured to give some account of the Metal it self.'[44] Boyle gives a separate account of his experiments on 'the filings of Zink or Spelter', not realizing that they were all (essentially) the same metal.

This new metal soon came to be highly prized, for it produced a far superior type of brass. The Germans sometimes gave it the name conterfe or conterfeht, since its appearance could rival that of gold.

In his Philosophical Principles of Universal Chemistry from 1730, German chemist Georg Ernst Stahl (1660–1734) compares the two methods of producing brass from copper, using either the minerals cadmia/calamine or the metal zinc:

> The other general Method of disguising or sophisticating Copper, regards the introduction of a yellow colour, whereby it is made to resemble Gold. And this is vulgarly effected by means of the Cadmia Plumbacea, Calamy, or Lapis Calaminaris; in the way of Cæmentation, or introducing it into the Metal by fusion. In which case 'tis remarkable, that the Calamy, tho' it be neither a compleat metallic body of it self, nor malleable; yet concretes along with the Copper, so as very considerably to increase its weight, and at the same time extend with it under the hammer. Whence the art of making Brass.
>
> Something of the same nature is likewise effected by Zink, tho' this gives the Copper a much more beautiful colour than the Calamy; and thus becomes the foundation of what they vulgarly call Bath or Prince's Metal, &c.[45]

Perhaps the first clear account for the unambiguous preparation of zinc metal is by the mining superintendent Georg Engelhardt von Löhneyss in 1617.

> When the people at the melting-houses are employed in melting, there is formed under the furnace, in the crevices of the wall, among the stones where it is not well plastered, a metal which is called zinc or conterfeht; and when the

wall is scraped, the metal falls down into a trough placed to receive it. This metal has a great resemblance to tin, but it is harder and less malleable, and rings like a small bell. It could be made also, if people would give themselves the trouble; but it is not much valued, and the servants and workmen only collect it when they are promised drink-money. They, however, scrape off more of it at one time than at another; for sometimes they collect two pounds, but at others not above two ounces. This metal, by itself, is of no use, as, like bismuth, it is not malleable; but when mixed with tin it renders it harder and more beautiful, like the English tin. This zinc or bismuth is in great request among the alchemists.[46]

The appearance of these crude pieces of zinc as jagged barbs of metal forming on the walls of the chimneys has suggested to some a possible etymology of the word zinc from the German words 'zacke' and 'zinke', meaning 'a jagged point or a prong'. But the first use of the word 'zinc' arises around the mid-sixteenth century, in the works of Paracelsus and Agricola. In his *Liber Mineralium II*, Paracelsus writes: 'Moreover there is another metal generally unknown called *zinken*. It is of peculiar nature and origin; many other metals adulterate it ... Its colour is different from other metals and does not resemble others in its growth.'[47]

Paracelsus regarded zinc as 'the bastard offspring of copper', and bismuth as that of tin. It has been suggested that such an association with one of the established seven metals may have given rise to the name used by Paracelsus, *zinken*. The German word for tin is 'zinn', and the suffix '-ken' could indicate a diminutive form (as in words such as 'manikin' and 'lambkin'). This suffix is related to the Germanic ending '-chen', used in such words as '*Mädchen*', 'a girl', and '*Kätzchen*', 'a kitten'). While this may or not be the origin of the word 'zinc', a similar route did give the name platinum. The metal was first found in South America as small silver-coloured nuggets that were called *platina*, meaning 'small silver' from the diminutive of the Spanish for 'silver', '*plata*'.

In the next chapter, we look at two elements that are associated with the gods, but also with the very depths of hell.

3

FIRE AND BRIMSTONE

This Sulphur from the Horrid deepe,
dame Nature did ordaine,
A fearefull scourge for sinne to be
as Scripture doth explane.

—Woodall, 1617[1]

The Element from Hell

Sulfur has long been associated with the fiery domain of hell, and with its god. In the fifteenth-century poem *The Assembly of Gods*, after describing Othea, the goddess of wisdom, the anonymous author continues with an account of the god of the underworld:

> And next to her was god Pluto set
> Wyth a derke myst envyroned all aboute
> His clothynge was made of a smoky net
> His colour was both wythin & wythoute
> Full derke & dȳme his eyen grete & stoute
> Of fyre & sulphure all his odour waas
> That wo was me while I behelde his faas[2]

Even more terrifying is the account from the Vatican Mythographers, in which Pluto is described as 'an intimidating personage sitting on a throne of sulphur, holding the sceptre of his realm in his right hand, and with his left strangling a soul'.

This association between sulfur and the fiery underworld is perhaps understandable given that the element is often found in the vicinity of volcanoes. In *Mundus Subterraneus*, one of many books written by the seventeenth-century polymath Athanasius Kircher (1602–80), the author describes a night-time visit to Vesuvius in the year 1638—just seven years after the great eruption of 1631. He tells us that after arriving at the crater, 'I saw what is horrible to be expressed, I saw it all over of a light fire, with an horrible combustion, and stench of Sulphur and burning Bitumen. Here forthwith being astonished at the unusual sight of the thing; Methoughts I beheld the habitation of Hell; wherein nothing else seemed to be much wanting, besides the horrid fantasms and apparitions of Devils.'[3]

Kircher believed that the volcanoes were fed by massive fires deep underground, as he tells us in the opening of his book:

> That there are Subterraneous Conservatories, and Treasuries of Fire (even as well, as there are of Water, and Air, &c.) and vast Abysses, and bottomless Gulphs in the Bowels and very Entrals of the Earth, stored therewith, no sober Philosopher can deny; If he do but consider the prodigious Vulcano's, or fire-belching Mountains; the eruptions of sulphurous fires not only out of the Earth, but also out of the very Sea; the multitude and variety of hot Baths every where occurring. And that they have their sourse and birth-place, not in the Air, not in the Water; nay, nor as the Vulgar perswade themselves, not at the bottom of the Mountains; but in the very in-most privy-Chambers, and retiring places of the Earth, is as reasonable to think; And there *Vulcan*, as it were, to have his Elaboratories, Shops, and Forges in the profoundest Bowels of Nature.

It is not surprising that some of the earliest theories of the origin of the subterranean heat assigned it to combustion; Kircher thought, 'The matter that doth nourish these Subterranean Fires, is Sulphure and Bitumen.'[4] Sulfur is certainly abundant near volcanoes. The master metallurgist Vannoccio Biringuccio (1480–ca. 1539), writing of sulfur in 1540 in his book *De la Pirotechnia* (*On the Fire-Crafts*), states that 'most liberal Nature makes whole mountains of it', and gives examples such as Mount Etna and the volcanic Aeolian Islands.

There was even some experimental evidence which seemed to support the idea that volcanic activity was due to the sulfur. At the very beginning of the eighteenth century, French chemist Nicolas Lemery famously created an 'artificial volcano' by burying in the ground fifty pounds of a mixture of iron filings and sulfur made into a damp paste. A repeat of the experiment, performed at Goodwood in the south of England in 1743, describes how, some eight hours after setup, the surrounding ground rises and shakes for an hour or two and then 'the Earth will swell, and heave, and burst at length with a Noise'. After this, blue flames appear and 'the Fire will last several Hours, and perfectly resemble that of natural Volcano's and Eruptions'.[5]

Subterranean fires do occur in nature—there are so-called burning mountains in both Germany and Australia, where coal seams have been smoulder-
ing for hundreds of years. The young poet Goethe visited the one in Dudweiler,
Germany in 1770 and wrote in his memoirs, '[A] strong smell of sulphur sur-
rounded us; one side of the cavity was almost red-hot, covered with reddish
stone burnt white; thick fumes arose from the crevices, and we felt the heat
of the ground through our strong boot-soles.'6 The vast coal-seam fires burning in the province of Xinjiang in north-west China cover an area of several
square kilometres and burn millions of tonnes of coal annually. Despite the
occurrence of these underground fires, we now know that combustion does
not explain the dramatic rise in temperature as one descends into the earth.
However, it was not until the very early twentieth century that the true origins
of geothermal heat were properly understood as being due to the radioactive
decay of certain long-lived isotopes, which adds a radiogenic contribution to
the primordial heat.

Heavenly Sulfur

Sulfur was not just confined to the underworld; it was also associated with
the thunderbolts unleashed from above by Pluto's brother and ruler of the
gods, Zeus. In Chapter 8 of Homer's *Iliad*, dating from the eighth century BC,
we find Zeus, with a powerful thunderclap, loosing his lightning and send-
ing a dazzling fiery bolt to hit the ground, where 'a dreadful flash came from
the blazing sulphur'. Similarly, in Chapter 15, Homer talks of Zeus using one
of his thunderbolts to destroy a mighty oak tree, which afterwards reeks
of sulfur.

Pliny also states that 'thunderbolts and lightnings in like manner doe sent
strongly of brimstone: the verie flashes and leames thereof stand much upon
the nature of sulphur, and yeeld the like light'.7 Even in the late seventeenth
and early eighteenth centuries, in addition to Lemery believing that sulfur was
the cause of subterranean fires and volcanoes, he also thought it explained
lightning and even hurricanes. In his course of chemistry from 1698, he writes:
'*Thunder*, therefore, ordinarily is produced by a sulphureous Wind, that is
enflamed and blown impetuously: Therefore, the places, where it passes, smell
strongly of *Sulphur*.'8 It is occasionally reported that there is a sulfurous smell
after lightning, but this is more likely to be due either to ozone (three oxygen
atoms united in a single molecule) or to oxides of nitrogen, formed during

the electrical discharge as the very molecules of the air are ripped apart and rearranged into new compounds.

'Brimstone' is the Old English for 'sulfur', and literally means 'burning stone'. Etymologists in the seventeenth century even suggested that the word had similar connotations, deriving from the Latin/Greek stems *sal* and *pyr*, signifying a fire-salt. It's more likely, though, that this ancient name is of Sanskrit origin.

The word 'brimstone' is used over a dozen times in the King James Bible. In Revelations we find several accounts of sinners and the devil being cast into a lake of fire and brimstone, where they 'shall be tormented day and night for ever and ever'. Similarly, it was brimstone and fire that the Lord rained down upon the sinful cities of Sodom and Gomorrah, a scene depicted in the *Nuremberg Chronicle* in 1493 (Figure 26), where we also see Lot and his daughters being led away by the angel and Lot's wife in her new guise as a surprisingly relaxed-looking pillar of salt, gazing back at the disintegrating cities.

The Greek word for 'sulfur', '$\theta\epsilon\acute{\iota}o\nu$' or '*thion*', is also the word meaning 'divinity', and perhaps it was the associations with fire-wielding gods that gave rise to this connection. In his *Compendious Body of Chymistry* from 1662, French chemist Nicaise le Fèvre (ca. 1610–1669) writes: '[I]t is not without reason that the Greeks gave to Brimstone the name of $\theta\epsilon\acute{\iota}o\nu$, that is to say Divine; for we must confesse that all Sulphurs have in themselves something heavenly and great...'[9] However, the connection could also be due to some of the other

Fig. 26. The destruction of Sodom and Gomorrah as depicted in a woodcut from the *Nuremberg Chronicle* of 1493.

properties of sulfur, notably its use in purification. When sulfur burns in air, it forms the choking gas sulfur dioxide, which has been used to fumigate houses, killing all insects and other pests (or, indeed, any living thing) with which it comes into contact. In Homer's *Odyssey*, after Odysseus returns and slaughters all the men who have taken over his home, he calls for sulfur and fire in order to purify the house. The Flemish physician and chemist Johannes Baptista van Helmont (1579–1644) tells how '*Hippocrates* named the hidden poyson of any diseases whatsoever, a divine thing', and because he had used sulfur to cure the pestilence, 'therefore he began to call Sulphur (τὸ θεῖον) [*thion*] that is, a divine thing; so that from hence even unto this day, Sulphur is no otherwise written or named, than with the name of Divine; because it heals the Pest'.[10] He then adds, 'The which, as it was antiently believed to be sent onely from the Gods, so also it was antiently supposed to contain a divine succour in it.'

The divine origin of sulfur remains with us in our modern chemical nomenclature. In compounds where an oxygen atom is replaced by a sulfur atom, the prefix '*thio-*' (or '*thi-*' if immediately before a vowel) appears in the name. For example, in thiosulfate, one of the four oxygen atoms in the sulfate ion, SO_4^{2-}, is replaced by a sulfur atom to give the thiosulfate ion, $S_2O_3^{2-}$. Similarly, ethanol, with the formula C_2H_5OH, is the alcohol present in wine, but ethanethiol has the oxygen replaced by sulfur and has the formula C_2H_5SH. Ethanol and ethanethiol have very different properties—we enjoy the former (in small quantities), but ethanethiol has an intolerable odour even when present at an extremely low concentration. Still, this disagreeable property has undoubtedly saved many lives since it alerts us to the presence of potentially explosive gas leaks—the thiol being added in minute amounts to the otherwise odourless natural gas.

Chemists say that the thiol unit is an example of a *functional group*—a particular arrangement of atoms that give compounds certain common properties whenever they occur in different molecules. An alcohol functional group may be denoted –OH (a hydrogen atom bonded to an oxygen) and the thiol functional group –SH (hydrogen bonded to sulfur). When the first thiol-containing molecules were prepared in the 1830s, they were initially called *mercaptans* by their discoverer, Danish chemist William Zeise; the ethanethiol mentioned above was therefore called 'ethyl mercaptan'. Zeise derived this name from the Latin '*corpus mercurium captans*', or 'mercury-capturing bodies', since the compounds readily reacted with some metallic preparations, notably with mercury oxide. Mercury has a strong affinity for sulfur—even today, mercury-spillage kits (for example, to deal with broken thermometers) contain powdered sulfur as the key ingredient which converts the metallic mercury into a safer form that is more easily contained. This reaction has been known for centuries. It is alluded to, for example, in the anthology of English alchemical poems, published in the seventeenth century as the *Theatrum Chemicum Britannicum*:

> I do liken our *Sulphur* to the Magnet Stone,
> That still draweth to her Naturally,
> So with our *Sulphur* the firey Woman Mercury,
> When she would from her husband flye.[11]

Natural sulfur such as that found near volcanoes would inevitably contain stones and earth and other impurities, and so needed refining. Biringuccio describes the equipment needed to distil the sulfur, and this is beautifully illustrated in Agricola's *De re metallica* from 1556 (Figure 27).

Fig. 27. A woodcut showing the distillation of sulfur from Agricola's *De re metallica*, 1556. The impure sulfur is heated in the pots (A) and the vapour from the boiling sulfur is fed into a larger pot (B), where it condenses to a liquid that drains out the spout at the bottom. The liquid may then be poured into moulds to form round cakes, or into wooden tubes moistened beforehand to make it easy to extract the rods of sulfur known as 'roll sulfur', shown in bundles on the floor.

Even from the time of Pliny, samples of sulfur that had been artificially worked and melted were thought to be different from naturally occurring samples. According to Agricola, native sulfur was known as *vivum* (living) or *apyron*, derived from the Greek meaning 'not exposed to the fire', as distinct from *pepyromenon*, 'exposed to fire'. However, sulfur could also be prepared as a fine powder by heating the sulfur in the absence of air and condensing the vapour on a cool surface. This form was known as *flowers of sulfur*, a term also occasionally used for the naturally occurring sulfur formed in a similar way. In Konrad Gesner's *The Newe Iewll of Health* from 1576, we find that the best sulfur is the 'sweating of the brimstone, which in brymstony places, out of hyls, as a flowre sendeth it forth: yet it may & ought to be named the flowre of the brimstone: for as ye dew, even so doth the sweate yssue forth of the stones'.[12]

Sulfur from Fire-stone

When native sulfur was not available, it could be extracted from sulfur-containing minerals such as pyrites. The name pyrites is now used in mineralogy for a yellow form of iron disulphide, better known as fools' gold, and has the chemical formula FeS_2. But before around the end of the eighteenth century, the term was used for a variety of minerals consisting of a metal, most often copper or iron, combined with sulfur. Agricola describes how sulfur could be extracted from this ore by carefully heating it (Figure 28).

The word '*pyrites*' is classical Latin and derives from the Greek '*pyrites lithos*', meaning 'fire-stone'. Agricola, through his characters Naevius and Bermannus, discusses the origin of the name in his earliest book from 1530:

Naevius: But Pliny writes, 'They call it pyrite because there is so much fire in it.' Is not fire obtained from it?

Bermannus: It is easy to strike fire from it and I believe, as Pliny, that the Greeks named it thus for this reason although it may have received this name because very often it is the colour of fire.[13]

The fire-generating properties of this mineral have been known since the earliest of times, when sparks would be struck off the pyrite using a flint. Samples of pyrites have been found with these so-called strike-a-light flints at many archaeological sites, such as the Star Carr site in North Yorkshire, which dates from around 8500 BC. Ötzi the Iceman, the 5300-year-old mummy found

Fig. 28. A woodcut from Agricola's *De re metallica* from 1556 showing the production of sulfur from pyrites. The ore is packed into a vessel with holes in its base (E) and this is placed on an iron plate with a hole in it (D). After lighting a fire around the vessel, molten sulfur is liberated, and it falls into pots of water placed below (F).

preserved in a glacier in the Alps on the border between Italy and Austria in 1991, had on his person an elaborate fire-lighting kit consisting of flint and iron pyrites together with various dried fungi and plants used as tinder. So the name 'pyrites'—or 'fire-stone', as it was sometimes also known in English—makes perfect sense for this mineral.

A more fanciful reason for the name is suggested in the medieval encyclopaedia *De proprietatibus rerum* (*On the Properties of Things*), written in the thirteenth century by Bartholomew the Englishman. He writes, 'Pirites is a redde bright stone, like to the qualitie of the aire: much fire is therein, and oft sparkles come out there of, and this stone burneth his hand that holdeth it right fast, therefore it hath that name of Pir, that is fire.'[14] This property of the stone burning the hand was noted earlier by Pliny, and later repeated by Albertus Magnus in his book on minerals. While at first it sounds highly improbable, it is possible that the idea originates from the fact that over time, in the presence of air and water, pyrites undergoes oxidation and can form strongly acidic solutions—in fact, the water in some mines can be dangerously acidic because of this reaction. Any acid present on the rock could, if not washed off, irritate and 'burn' the hand. This reaction was well known to Agricola, who writes: 'experiment shows that when porous, friable pyrite is attacked by moisture such an acid juice is produced'.[15] As well as taking place naturally, the process could be carried out artificially, and this has led to one of the most important uses of both sulfur and pyrites: the production of sulfuric acid.

Shoemakers' Black, Copperas, and Vitriol

In addition to the acid formed when metal sulfides such as iron or copper pyrites react with oxygen and water, water-soluble metal sulfates are also produced. Agricola describes how the sulfate crystals sometimes form around the pyrite, so that 'in the centre of these masses pale-coloured pyrite is found almost dissolved'.[16] He adds that the solutions may 'come out from the rock drop by drop and, moving down along channels, congeals in the form of icicles…which the Greeks call σταλακτικός [stalaktikos] because it has congealed by dropping'. While our word 'stalactite' indeed derives from the Greek word 'stalaktos', meaning 'dripping or dropping', the examples we commonly see in caves are most often made of calcium carbonate. Technically, any soluble substance could form similar structures. Sulfates of both copper and iron were also familiar to Pliny, although the two were often muddled, even though the pure forms have different colours (green for iron sulfate and blue for copper). The Latin term used for 'iron sulfate' was 'atramentum sutorium', which literally means 'shoemakers' blackening'. This curious name arose since solutions of iron sulfate give a black pigment with the tannins found in leather, and so were much used in that trade. Although the sulfates of copper and iron share some similar properties, only iron sulfate forms the black colour with tannins.

Pliny also tells us that the Greeks referred to the sulfates as *chalcanthos*, which literally means 'copper flower', because of the beautiful growth of the sulfate crystals which forms on certain minerals. This same connection has also been suggested for the term 'copperas', sometimes written 'coperose' or 'cupri rosa', 'rose of copper'. To make things really confusing, 'copperas' was later used more for green iron sulfate than for blue copper sulfate. More likely than the floral origin, though, is the idea that the term was simply short for the Latin '*aqua cuprosa*', meaning 'cuprous water'. Even so, this cuprous water might still have referred to iron sulfate and its solutions. In the Latin edition of his *De re metallica* from 1556, Agricola uses the term '*atramentum sutorium*' (the shoe-blackening iron sulfate), but in the German edition from the following year, it becomes '*Kupfferwasser*' (literally, 'copper water').

Pliny also describes how crystals may be prepared from solutions of the sulfates by hanging into them ropes with little stones tied at the ends. The crystals form on them as glassy berries, not unlike grapes. He adds that the dried material 'is blue, with a very notable brilliance, and may be mistaken for glass'.[17] In his book on minerals written in the thirteenth century, Albertus Magnus states that some people call the green kind *vitreolum* (glassy)[18] and the term '*vitriol*' became a common term for the sulfate crystals, being used, for example, by Biringuccio in 1540. Biringuccio gives a detailed account of the preparation of vitriol (iron sulfate) in which the crude sulfates (a mixture of both copper and iron sulfates) are dissolved in hot water, and then scraps of iron are added. During the reaction, copper metal deposits on the iron, and the iron itself gradually dissolves into the solution. What is left is a much purer solution of green iron sulfate.

Oil of Vitriol and Spirit of Sulfur by the Bell

One of the most important uses of vitriol was to make sulfuric acid. Simply heating the iron sulfate first drives out water—the so-called water of crystallization contained in the crystals. At higher temperatures, the sulfate breaks down and volatile oxides of sulfur are driven out, as described in Gesner's *The Newe Iewell of Health* from 1576: 'you shall then see the spyrites [spirits] yssew forthe, even lyke to cloudes heaped togither'.[19] The sulfur oxides dissolve in water to form the acid, and all that remains in the heated vessel is rust-red iron oxide. Two oxides of sulfur are formed during this process: gaseous sulfur dioxide, SO_2, and a volatile solid, sulfur trioxide, SO_3. The latter reacts with water to form sulfuric acid, but the former gives an unstable solution known as sulfurous acid. Over time, when exposed to the oxygen from the air, sulfurous acid oxidizes to the stronger sulfuric acid. Sometimes a distinction was made between the two acids—'oil of vitriol' for the more syrupy sulfuric acid, and 'spirit of sulfur' for the milder (although more choking) sulfurous acid.

Another way to generate the acids was directly from sulfur. While heating the sulfur in the absence of air distils the sulfur and forms flowers of sulfur,

Fig. 29. The preparation of 'sulfur by the bell'. The woodcut on the left is from Gesner's *The Newe Iewell of Health* from 1576. The engraving on the right is from a text from 1690.

when air is present, the sulfur burns with a sinister blue flame, forming mainly sulfur dioxide (the reaction used by the Greeks in purification, as we saw earlier). Recipes describe how to prepare the so-called volatile spirit of sulfur using an apparatus consisting of a glass or ceramic 'bell' designed to condense the spirit, held above the burning sulfur so as to allow plenty of air to get to the sulfur (Figure 29).

The procedure is rather odd in that the sulfur dioxide, formed as the burning sulfur combines with the oxygen from the air, is a gas at room temperature and so would not condense on the glass bell—even less so as the bell became warmed by the flame. A clue as to how the process might work at all is provided in le Fèvre's *Compleat Body of Chymistry* from 1664, where he comments on the best time of year to carry out the operation: 'above all times chuse that of the two Æquinoxes, vernal and autumnal, to work this Spirit. That season being moist for the most part and rainy, which is a thing necessary in this operation, otherwise you shall draw very little spirit from lib. j. [one pound] of Brimstone'.[20] He adds that 'if the ayr be too dry by intervention of either cold or heat, it is not capable of coagulating the acid and vitriolick spirit of the Brimstone, which contrariwise is totally dissipated with the fat and inflammable substance of the Brimstone'. The sulfur dioxide readily dissolved in any moisture to form sulfurous acid, and some chemists sensibly chose to moisten the bell with water beforehand. Nonetheless, however it was carried out, the process was extremely inefficient, and Gesner noted 'that of fyve poundes of Brimstone, you shall hardly gather one ounce of oyle'.[21] The 'oyle' was known as *sulphur per campanam* or 'sulfur by the bell', reflecting its method of production.

A significant improvement came about with the addition of nitre, or potassium nitrate, to the sulfur. Perhaps this was initially added to stop the flame of the sulfur going out, but it turns out that it provides another function. As the nitrate decomposes, oxides of nitrogen can be produced which help to catalyze the reaction between sulfur dioxide and the oxygen from the air to form sulfur trioxide. A quack doctor, Joshua Ward, set up a works in Twickenham, London, to make the oil of vitriol necessary for his dubious remedies. He used enormous glass globes, 60–70 cm in diameter with wide necks, to replace the earlier bell, with the sulfur/nitre mix placed in the centre above a little water. Just how vast these vessels were is brought home by Fellow of the College of Physicians Dr Samuel Musgrave, who reported on their manufacture: 'Nothing however requires a longer expiration than the blowing large glasses with the blow-pipe. I have been told that in blowing the large glass recipients, in which the late Dr Ward used to collect the spirit of vitriol, it was not uncommon for the blood to start out forcibly from the nose and ears of the person employed to blow them.'[22] Others learnt that the troublesome glass vessels could be more conveniently replaced by lead containers, and soon the reaction was carried out in massive lead-lined rooms or 'chambers', some 3 m square and 13 m long. With further refinements, this so-called lead-chamber process continued to be used for the production of sulfuric acid well into the twentieth century.

On 18 November 1731, one Sigismund Augustus Frobenius, a German-born chemist living at the time in London, exhibited before the Royal Society 'a very pompous Machine, which he calls *Machina Frobeniana, pro resolutione Combustibilium*' (Frobenius's machine for resolving by combustion).[23] It was noted that this machine was really no more than a fancy version of the apparatus for producing the oil of sulfur by the bell, and the demonstration was repeated straight afterwards by another chemist, Ambrose Godfrey Hanckewitz, using a glass jar and a warm china cup. But rather than using sulfur in the bell, the chemists were burning a new substance—one of the most reactive of all the elements—called in the report '*Phosphorus glacialis Urinæ, or Stick Phosphorus of Mr Ambrose Godfrey Hanckewitz*'.

The Morning Star, Bearer of Light

The term 'phosphorus' dates back to the ancient Greeks, who used it to refer to the planet Venus when visible in the eastern morning sky before dawn. With its literal meaning of 'bearer of light', Phosphorus was an apt name for the brightest object in the sky after the Sun and Moon; in his play *Ion* from around 414 BC, Euripides writes, 'light-bearing Dawn [Phosphorus] puts the stars to flight'. Curiously, the Greeks had a different name—Hesperos—for the planet Venus when visible in the evening before sunset; perhaps since at one time it was not appreciated that the Morning and Evening Star were the same body. In Greek mythology, both Hesperos and Phosphorus were sons of the goddess of the dawn, Eos (her Roman equivalent being Aurora), explaining the alternative name sometimes used for this meaning of phosphorus: Eosphoros—the Bearer of Dawn. The Latin equivalent of 'Phosphorus' is 'Lucifer', and it was in translating the Hebrew for 'Morning Star' as 'Lucifer' that the association of this name with Satan arose. However, to understand why the name phosphorus was used for the devilish element, we need to go back over half a century before its discovery, to the beginning of the seventeenth century in rural Italy.

The Solar Sponge

In 1602, a humble cobbler from Bologna, Vincenzo Cascariolo, discovered that after calcining or roasting in his furnace certain stones found on the slopes of the nearby Monte Paterno, they had the remarkable property of being able to absorb light and then emit it as an eerie glow in the dark. The English naturalist John Ray (1627–1705), who later related tales of his travels on the continent, described visiting 'Seignior Gioseppi Bucemi a Chymist' in Bologna, who prepared the stone 'which if exposed a while to the illuminated air will imbibe the light, so that withdrawn into a dark room, and there look't upon it will appear like a burning coal'. He adds that it 'in a short time gradually loses its shining till again exposed to the light'.[24]

We would now say that Cascariolo had achieved the first preparation of a persistent luminescent or phosphorescent material, modern improvements of which are used in emergency exit signs or children's toys that glow in the dark after absorbing energy from light. Experiments from 2012 have shown the Bolognian stone to be impure barium sulfide, formed from natural samples of heavy spar, or barium sulfate. The key thing in the Bolognian stone are the trace impurities, notably singly-charged copper ions. During exposure to light, electrons in the copper ions become energetically excited and trapped in defects in the barium sulfide crystal. Over time, the electrons return to their lower-energy state, emitting the stored energy as light once again. Of course, Cascariolo knew nothing of this; he simply thought that his stone 'is accustomed to imbibe the light of the Sun as the sponge soaks up liquid' and he therefore called it *spongia solis* ('sponge of the sun' or 'sponge of sunlight'). Athanasius Kircher, whom we met earlier in the chapter descending into Vesuvius, wrote in 1641 that the stone was a kind of magnet acting on light in the same way that an ordinary magnet acts on pieces of iron. It was even suggested that moonlight might be partly due to the light emitted from a similar substance, rather than being the reflected light of the Sun. Galileo, who was also familiar with the Bolognian stone, vehemently opposed this idea. As fame of the wondrous stone spread and it was realized that it could be charged not only with the light of the Sun but also by the Moon, and even the light from a flame, it received a plethora of names such as '*spongia lucis*' ('sponge of light'), '*retinaculum luminis caelestis*' ('holder of heavenly light'), '*lapis illuminabilis*' ('the stone that can take on light'), '*lapis lucifer*', and '*lapis phosphorus*' (both meaning 'the stone that carries light').

The Bolognian stone was notoriously difficult to prepare—Ray reported that 'there is somewhat more of mystery in it; for some of us calcining part of the stone we purchased of him according to his direction [laying the pieces of stone upon an iron grate over a fire of wood], it sorted not to make it shine'.[25] In the 1660s, it even seemed that the art of preparing the stone was lost—but then a German chemist, Wilhelm Homberg, learned the secret after travelling to Italy. Homberg soon had a better understanding of the stone than anyone else had, and he prepared excellent phosphorescent stones in Italy; but when he arrived in Paris, where he eventually settled after extensive travels around Europe, he found, much to his annoyance, that, despite his many attempts, the procedure no longer worked. Then a chance encounter provided the key. He had promised to a friend that he would tell him how to prepare the stone, not having confessed that the method no longer worked. After attempting to postpone the demonstration, he happened to bump into the friend, who marched him off to his home laboratory in order to be shown exactly how to prepare the marvel. The friend had already built a furnace following Homberg's instructions, and had the raw, uncalcined stones that Homberg had given him earlier. Homberg later wrote of the encounter: 'Being thus pressed, I began again the operation which had so often failed, and to speak the truth, I was trembling all the while, for I had not told him that I had always failed at it in Paris.'[26] But to his surprise and relief, he continues, 'When the operation was finished I found the stones the most brilliant and luminous that I had ever seen.' After going through what he had done differently, he finally realized the key factor: in his own lab in Paris, he had used an iron mortar to grind the raw stone, but in his friend's, and when in Italy, he had used a bronze mortar. After an extensive series of trials, he then realized the secret—grinding in copper or bronze greatly enhanced the luminescence of the stones, but the presence of iron inhibited it. We now know that as well as copper ions being crucial for the stone to absorb and re-emit light, the presence of iron destroys the phenomenon—it is said to quench the phosphorescence. In 2016, historian of chemistry Lawrence M. Principe published an excellent account of the stone, including his experiments to prepare it.[27] Principe found not only that the presence of copper and absence of iron were crucial, but also the precise way in which the furnace was constructed. The furnace type described by Homberg actually meant that gaseous carbon monoxide acted on the barium sulfate during the calcination to reduce it to the sulfide. Simply heating the raw stone with charcoal would not produce a phosphorescent stone.

The Light Magnet

For seventy years after its discovery, the Bolognian Stone was a unique marvel of chemistry, but eventually other substances began to be discovered that could also glow in the dark. One of these was first prepared with the intention of being used to attract moisture out of the air. The water, once extracted and purified, could be sold for a high price as a quack medicine. We'll come across this substance again later, but after an accident during its preparation, its discoverer, Christian Adolph Balduin, soon found out that his 'magnet' attracted more than just the water from the air. Balduin prepared what we would now call calcium nitrate by dissolving chalk (calcium carbonate) in nitric acid and evaporating the resulting solution to leave the solid. Johann Kunckel (1630–1702), who we shall see plays a key role in the discovery of phosphorus, reports Balduin's famous discovery in his posthumously published *Laboratorium Chymicum* from 1716[28]: 'In the course of this work it happened by mistake that the spirit of niter was once evaporated to a hard mass, and consequently that something yellow collected in the neck of the retort. After it had been broken indoors, he threw the neck into a dark corner of the laboratory where he remarked that it glowed like a coal. He observed the phenomenon with wonderment, and remarked that this light faded again in the dark and took on light again from the sunlight.'[29] Balduin published the account of his new wonder in 1675 in an appendix entitled 'Phosphorus Hermeticus, sive Magnes Luminaris', which may be translated as 'The Hermetic Phosphor [light-bearer] or Light Magnet'. His remarkable substance soon came to be known simply as Balduin's Phosphorus. Despite being called phosphorus, Balduin's substance contains none of the element we now know by this name; the name was simply used in its capacity as meaning 'light-bearer'. Balduin's preparation did, though, give rise to the term 'phosphorescence'—the process by which a substance absorbs light energy and then re-emits it later. Ironically, the element phosphorus does not do this. The light it generates is from a chemical reaction with the air, in which the phosphorus is gradually used up—a different process, called chemiluminescence.

After first hearing of Balduin's Phosphorus, Kunckel was keen to learn what it was and how to prepare it. He visited its discoverer, whom he found extremely hospitable but not at all forthcoming in revealing his secret; Kunckel states that 'his discourse was as orderly as a swarm of bees'. When Balduin left the room to find a concave mirror to enhance the light falling on the substance, in his haste, he accidentally left the material unattended; Kunckel seized the opportunity and 'twitched off a little and put it into my mouth'.[30] Perhaps Kunckel's suspicions of the composition of the phosphor were confirmed on tasting the substance, for he was immediately able to produce his own version, much to the irritation of Balduin.

It was while exhibiting the magical properties of Balduin's substance to an audience in Hamburg in the 1670s that Kunckel first came to learn of another that was even more amazing. A member of the audience, a preacher named Peter Hessel, approached him and said, 'There is a man here, called Doctor Brand, an unsuccessful merchant who has applied himself to medicine, who has recently made something which glows continuously in the night.'[31] The fact that this substance glowed continuously in the dark was immensely striking since up to this point, all the other light-bearers needed to be charged up by exposure to light, and then gradually faded over hours, or even minutes.

Cold Fire

Phosphorus is the earliest element for which both the year of discovery (1669) and the name of its discoverer (Hennig Brand from Hamburg) are known.

Of course, it was not appreciated that it was an element at the time—this realization came over a century later. Kunckel was taken to meet Brand, who showed him what he called 'cold fire' or just 'my fire'. Brand did not publish any account of his discovery, and it only became more widely known as a result of Kunckel's visit. Kunckel was desperate to learn from Brand how he could prepare the phosphorus himself, but before he had managed to secure the procedure, Kunckel made the mistake of writing about the discovery to one of his friends, Johann Daniel Krafft (or Crafft). Krafft wasted no time and immediately came to Hamburg and bought the secret for himself, and even paid Brand not to inform Kunckel. After much fruitless correspondence with Brand, it seems Kunckel eventually managed to prepare phosphorus through his own skill, and in 1678, he published a book whose title may be translated as *Open Letter on the Phosphorus Mirabilis and on Its Glowing Wonder Pills*. Much of the phosphorus described was actually rather impure, with the appearance of a 'black soap' in which particles 'flash and twinkle like little stars'. Kunckel notes that when it is rubbed into hair, 'each hair gives off a glow—and, once seen, is a thing to be remembered'.[32] He also describes writing with it, but warns 'if one presses too strongly, the paper takes fire'. We now know that this form of phosphorus prepared by the first investigators is extremely toxic (more so than cyanide), but Kunckel made his 'Wonder-Pills' by allowing the phosphorus to stand in solutions of gold or silver, allowing a chemical reaction to occur which doubtless saved the lives of his patients. In this reaction, the phosphorus is converted into non-toxic phosphoric acid (the acid ingredient in fizzy cola drinks), and an attractive coat of metal forms in place of the poison. Kunckel remarks, 'They cause no vomiting nor any inconvenience but act in a mysterious manner and are applicable in serious sickness and pain…'. He even recommends giving them 'to little children of a few weeks of age who frequently cry night and day and have no repose'.[33]

Kunkel's work on phosphorus came out in 1678, but it was not the first publication on this new substance. That was a little pamphlet that had appeared in Berlin in May 1676, after Krafft had exhibited at the court of the Grand Elector Friedrich Wilhelm von Brandenburg the miraculous material he had obtained from Brand. The title of the book is *De phosphoris quatuor, observatio*, and it describes the four glowing substances then known: the Bolognian Stone, Balduin's Phosphorus, a recently discovered variety of fluorite which glowed when warmed (see Chapter 7), and finally the new phosphorus, called *phosphorus fulgurans*, or 'flashing phosphor'. The latest substance is enthusiastically described: 'All the previously named species of phosphors are left far behind by the fourth and most recent, for which the name "flashing phosphor" is fitting because of its special activity.'[34] During his demonstrations, Krafft called the substance 'eternal fire' ('*Ignem perpetuum*'). The report states that the phosphorus 'not only lit up itself as do the glowworms that fly through the air on summer nights, but to the astonishment of the onlookers it also transferred the same whitish shimmer to the finger with which it had been rubbed. If anyone had rubbed himself all over with it, his whole figure would have shone, as once did that of Moses when he came down from Mt Sinai (if the comparison with such sacred matters is permissible here).'

In a letter to Brand dated 25 June 1676, Kunckel writes, 'Krafft and I, although hitherto close friends, have almost become enemies over this, because he boasted so brazenly at Berlin, and permitted a physician to print a pamphlet about it spreading the impression that the discovery is really Krafft's. I have refuted this.'[35] Kunckel may have refuted Krafft as the discoverer, but he did not acknowledge Brand in his own book, and seemed happy for others to assume he was the discoverer. Consequently, the substance frequently came to be called Kunckel's Phosphorus.

While the early publications report Krafft's and Kunckel's spectacular demonstrations of the new substance, they were careful not to mention any hint of how the phosphorus was actually prepared. The first recipe appeared in 1680, after Robert Boyle became fascinated with the wonder.

Phosphorus Comes to London

On Saturday 15 September 1677, Krafft exhibited the 'strange rarity' at the London house of Robert Boyle, who later published a report of the event.[36] Krafft unpacked a number of glass vessels, the largest being a sphere four or five inches in diameter containing a couple of spoonfuls of what looked like muddy water; a few tubes; and a small button-bottle containing 'a little lump of matter…that appeared of a whitish colour, and seemed not to exceed a couple of ordinary Pease, or the kernel of a Hasel Nut in bigness'. With everything laid out on the table, 'the windows were closed with woodenshuts, and the Candles were removed', leaving everyone in the dark. 'Though I noted above that the hollow Sphere of Glass had in it but about two Spoonfuls (or three at most) of matter, yet the whole Sphere was illuminated by it, so that it seemed to be not unlike a Cannon bullet taken red hot out of the fire, except that the light of our Sphere lookt somewhat more pale and faint.' When he held the sphere in his hands, Boyle noted the liquid appeared to shine more vividly, and sometimes flashed. Krafft then took a small piece of solid phosphorus and crushed it into tiny fragments and 'scattered them without any order about the Carpet, where it was very delightful to see how vividly they shined'. Boyle thought 'they seemed like fixt Stars…and these twinkling sparks without doing any harm (that we took notice of) to the Turky Carpet they lay on, continued to shine for a good while'.

'Mr *Kraft* also calling for a sheet of Paper and taking some of his stuff upon the tip of his finger, writ in large Characters two or three words, whereof one being *DOMINI*, was made up of Capital Letters, which being large enough to reach from one side of the page to the other, and being (at least as I guessed) invigorated by the free contact of the external Air, shone so briskly and lookt so oddly, that the sight was extreamly pleasing, having in it a mixture of strangeness, beauty and frightfulness, wherein yet the last of those qualities was far from being predominant.' Boyle also noted the characteristic 'odour of Sulphur and of that of Onions' associated with the phosphorus.

Finally, Krafft took some of the luminous matter and rubbed it upon the back of Boyle's hand and on his cuff: 'And all this while this light that was so permanent, was yet so mild and innocent that in that part of my hand where it was largely enough spread, I felt no sensible heat produced by it.'[37]

Boyle's Aerial Noctiluca

Spurred on by Krafft's demonstrations, Boyle wanted to prepare his own phosphorus. Krafft would not tell Boyle how the substance was prepared, but after Boyle revealed to him an alchemical secret, 'he, in requital, confest to me at parting, that at least the principal matter of his *Phosphorus's*, was somewhat that belong'd to the Body of Man'.[38] Boyle suspected this must mean the product was obtained from urine and so immediately set to work using materials he had to hand from earlier experiments. Despite many trials, he could not obtain any of the desired product. He was then given a hint by a 'learned and ingenious Stranger, (A. G. M. D. Countreyman, if I mistake not, to Mr *Krafft*)'. It has been suggested that the stranger was Ambrose Godfrey (Hanckwitz), who later became Boyle's assistant and ultimately the leading supplier of phosphorus to all of Europe. Whoever the stranger was, the crucial clue provided was that the process depended on 'the degree of Fire'. It turns out that extremely high temperatures are needed in the preparation, and this is why many attempts had failed. The original discoverer, Brand, and Kunckel (who independently learned how to prepare phosphorus) had both been glass workers at one time, and so were used to preparing and using high-temperature furnaces. Boyle set his assistant to work once again distilling the foul residues into the receiving vessel. This time he was so confident it would work that, 'I would not believe the skilful *Laborant*, when he told me with trouble, that what I expected, was not at all produc'd: But going my self to the *Laboratory*, I quickly found, that by the help of the *Air*, or some *Agitation* of what had pass'd into the *Receiver*, I could, in a dark place (though it was then day) perceive some glimerings of light, which, you will easily believe, I was not ill pleas'd to see.'

Unlike the other discoverers of phosphorus, Boyle set about publishing detailed accounts of his experiments and how to prepare the substance. His book, published in 1680, first describes the other phosphors of the day, namely the Bolognian Stone and Balduin's Phosphorus. He then writes: 'There is *another* sort, which needs not be previously illustrated by any external Lucid, and yet continues to shine far longer than the *Bolonian Stone*, or the *Phosphorus* of *Balduinus*. This, by some Learned Men has been call'd, to discriminate it from the former, a *Noctiluca*.'[39] The word 'noctiluca' derives from the Latin words '*nox*' ('light') and '*lucere*' ('to shine') and was used to imply that this substance, unlike the other phosphors then known, was capable of shining at night without first being illuminated; although Boyle points out that 'in strictness I cannot think it as proper a name as could be wish'd, since the other *Phosphorus* will shine in the Night as well as the Day, if it be excited with the flame of a culinary Fire, or of a large Candle'. Despite not liking the word 'noctiluca', Boyle uses it, but also substitutes for it the term '*Self-shining substance*, which is more expressive of its nature'. Boyle notes that Krafft had earlier shown His Majesty King Charles II two sorts of phosphorus, a liquid and a waxy solid—the latter Boyle called a 'Gummous Noctiluca' or 'Consistent Noctiluca'. He also noted 'that on the score of its uninterrupted action, 'tis call'd by some in *Germany, The Constant Noctiluca*; which title it does not ill deserve, since this *Phosphorus* is much the *noblest* we have yet seen'.

The phosphorus Boyle first prepared was far from pure, and his initial attempts yielded only a liquid which did not itself glow but seemed to cause the vapour above it to shine when air was admitted to the vial. He writes: 'the Substance that shin'd, was not the *Body of the Liquor* included in the Vial, but an Exhalation or *Effluvium* mingled with the admitted Air: for both which Reasons, I gave it the name of *Aerial Noctiluca*'. This was the title of his book—*The Aerial Noctiluca*.

In his book, Boyle highlights the confusion over who actually first discovered the new substance. He writes: 'For though I find it generally agreed, that the *Phosphorus Hermeticus* was first found and published to the World, by the learned and ingenious *Balduinus*, a *German* Lawyer; yet as to the *Gummous* and *Liquid Noctiluca's*, I find the first invention is by some ascrib'd to the abovemention'd Mr *Krafft*, (though I remember not, that when he was *here*, he plainly asserted it to himself;) by others, attributed to an ancient Chymist, dwelling at *Hamburgh*, whose name (if I mistake not) is Mr *Branc*, and by others again, with great confidence, asserted to a famous *German* Chymist in the Court of *Saxony*, call'd *Kunckelius*. But to which of these so Noble an Invention, as that of the two *German Noctiluca's*, is justly due, I neither am qualified nor desirous to judge.'[40]

The Recipe

In his *Aerial Noctiluca*, Boyle gives the very first account of how phosphorus may be prepared by heating at very high temperatures the residues formed after distilling concentrated human urine:

> There was taken a considerable quantity of *Humane Urine*, (because the Liquor yields but a *small* proportion of *luciferous matter*,) that had been, (a good part of it at least) for a competent while, digested or putrified, before it was us'd. This *Liquor* was distill'd, with a moderate heat, till the *spirituous* parts were drawn off; after which, the superfluous *moisture* also was abstracted, (or evaporated away) till the remaining substance was brought to the consistence of a some-what *thick syrup*, or a *thin extract*. This was well incorporated with about thrice its weight of *fine white sand*, and the mixture was put into a strong *Retort*; to which was join'd a *large Receiver*, in good part fill'd with water. Then, the *two* Vessels being carefully luted together, a naked Fire was gradually administred, for *five* or *six* hours, that all, that was either *Phlegmatick*, or otherwise *Volatile*, might come over *first*. When this was done, the Fire was increas'd, and at length, for *five* or *six* hours made (NB) which it should be in this Operation) as strong and intense, as the *Furnace* (which was not bad) was capable of giving. By this means, there came over good store of white fumes, almost like *those*, that appear in the Distillation of *Oil* of *Vitriol*; and when *those* fumes were past, and the *Receiver* grew clear, they were after a while succeeded by *another* sort, that seem'd in the *Receiver* to give a faint *blewish light*, almost like *that* of little burning Matches, dipt in *Sulphur*. And last of all, the Fire being very vehement, there pass'd over *another* substance, that was judg'd more ponderous than the former, because (NB) much of it fell through the *water* to the bottom of the *Receiver*: whence being taken out, (and partly even whil'st it staid *there*) it

appear'd by several effects, and other *Phoenomena*, to be (as we expected) of a *luciferous nature*.[41]

Boyle is rather vague about the 'considerable quantity' of urine used, but another key player in this saga, Gottfried Wilhelm Leibniz, the great mathematician and polymath who discovered calculus independently of Isaac Newton, wrote a similar recipe in a letter in 1682. This recipe, which Leibniz almost certainly got directly from Hennig Brand himself, begins: 'Take approximately a full ton of urine that has stood for some time…'[42] Furthermore, at one time Brand was employed by Leibniz and his Patrons to produce phosphorus on a large scale using human urine provided from a garrison of soldiers; this is said to have involved 100 tonnes of urine, corresponding to 13,140 litres.

Two years after the publication of the *Aerial Noctiluca*, Boyle published a follow-up work, *New Experiments, and Observations, Made upon the Icy Noctiluca*. In this he describes the properties of much purer, solid phosphorus. Regarding the name for the solid lumps, he writes: 'And some of the bigger appeared so like such Fragments of *Ice*, as being thin, are oftentimes very clear, and almost quite destitute of manifest Bubbles; that because of this great resemblance, and for distinction sake, I thought it not amiss to call our consistent Self-shining Substance, the *Icy* or *Glacial Noctiluca* (and for variety *Phosphorus*).'[43]

Boyle found that his experiments were less acceptable 'to the delicate sort of Spectators, especially to Ladies' because of the unpleasant smell that accompanied the phosphorus. To get around this, he tried dissolving the phosphorus in various aromatic oils; the solution in oil of cloves gave a strong light 'far more vivid…than any Liquor had afforded us before'.

Boyle also reports how the phosphorus can cause painful burns: 'If our *Phosphorus* be for any time pressed hard between ones Fingers…it will oftentimes be felt actually and very sensibly hot, and sometimes the degree of heat will be so vehement, as to Scorch the Skin, as my venturous Laborant found several times to his no small pain, his Fingers being almost covered with Blisters raised on them, by handling our shining Matter, with too bold a curiosity.'[44] Phosphorus burns are notoriously painful, and Boyle adds that his assistant 'complained to me, that, though according to the usual fate of *Chymists*, he had been often Burned on other occasions, yet he found Blisters, excited by the *Phosphorus*, more painful than others; and he is not the only person that has complained to me of their finding the Burning made with this Matter to be more tedious and difficult to be cured, than ordinary ones.' Boyle's poor assistant suffered several mishaps with the dangerous element. On another occasion, Boyle was trying to light gunpowder with a small piece of phosphorus. At first, nothing seemed to happen, but then his hapless assistant leant too far over the mixture 'but then upon a sudden the powder took Fire, and the flame shooting up, caught hold of his Hair, which made a Blaze, that proving innocent enough, became more diverting, than the smell of the Smoke that succeeded it was delightful'. Boyle adds that the same worker had a worse misadventure not long afterwards, when a bottle of phosphorus he was carrying in his pocket broke. The substance 'Burned two or three great holes in his Breeches, before he could come to me to relate his misfortune, the recent effects of which I could not look upon without some wonder as well as smiles.'[45]

Despite all these painful accidents, people could not resist touching the poisonous wonder. In *Chymicus Rationalis*, a short text by William Y-Worth from 1692, after describing the preparation of 'Fosperus', we learn that 'if rubbed upon the Hands, Cloaths, or Hair, they will appear in the dark, as if all in fire, but will not burn'; most disturbingly, it then adds 'If the Privy Parts be therewith rubb'd, they will be inflamed and burning for a good while after.'[46]

Ambrose Godfrey (he later essentially dropped the Hanckewitz) became famous for, and wealthy from, producing the best phosphorus in Europe. He writes in 1731, 'And I know my self to have been for these forty or fifty Years, that is, ever since I left the Laboratory of my Master the Honourable Mr *Boyle*, the only Person in *Europe* able to make and produce in any Quantity the true solid *Phosphorus*.'[47] So famous was his product that it was often referred to as 'the English Phosphorus'. Godfrey used not only urine to make the phosphorus, but also faeces. In the paper reporting the demonstrations he made with Frobenius before the Royal Society, Godfrey describes the disgusting lengths he went to in his trials:

I did not content my self to work upon the *Urinous Sapo* of Man only, but examined likewise the Excrements of other Animals; as for Example, of Horses, Cows, Sheep, &c. and got *Phosphorus*, but not in so great Quantities as from Man; probably because they feed on nothing but Vegetables. I then examin'd the Dens of Lions, Tygers, and Bears, making Experiments on their Excrements, and likewise on those of Cats and Dogs, which being carnivorous Animals, I obtain'd more *Phosphorus* thence than from the other Creatures: My Curiosity led me likewise toe the Rats-Nests, and Mouse-Holes, and I had *Phosphorus* thence. I then address'd my self to the feather'd Tribe, visiting the Hen-Roosts, and Pidgeon-Houses, and got some small Matters thence also: I emptied the Guts of Fish in order to get their Excrements, and had a little *Phosphorus* from these, but none from the Fishes by themselves.[48]

Thankfully, ways were found later in the eighteenth century to prepare phosphorus from other sources—first from bones, and later from phosphate minerals.

All of the preparations of phosphorus so far discussed are a form we would now call 'white phosphorus'. In the 1840s, a new and much safer form was discovered—it is not a deadly poison, nor is it spontaneously flammable in air. This variety, formed by heating white phosphorus for extended periods of time, is called 'red phosphorus'. Both are forms of the element phosphorus not combined with any other element; they differ only in how the atoms of phosphorus are arranged. The white variety is made up of P_4 molecules—four phosphorus atoms bonded to each other to form individual tetrahedral units. In red phosphorus, countless phosphorus atoms join up to form extended networks. There are other forms too, also with different colours, including black and violet.

Sales of white phosphorus are now prohibited since it is so dangerous and easily used in weapons. In a tragic twist of fate, Brand's home city of Hamburg, where the wondrous light-bearer was first discovered, was destroyed during the Second World War, partly by bombs containing white phosphorus. The Battle of Hamburg, codenamed Operation Gomorrah, started on 24 July 1943 and led to the near-total destruction of the city and the loss of tens of thousands of civilian lives. Sadly, white phosphorus has also been used aggressively in the twenty-first century.

4

'H TWO O' TO 'O TWO H'

'[W]e do not hesitate to conclude that water is not a simple substance, and that it is composed, weight for weight, of inflammable air and vital air.'
—Lavoisier, 1784[1]

It was not until the late eighteenth century—over a hundred years after the discovery of phosphorus—that it was appreciated that both phosphorus and sulfur were actually elements. Prior to this time, it was thought that all matter was made up of four so-called elements: earth, air, fire, and water. The realization that this was not so centred on understanding that the air is actually composed of a number of different gases, and in particular, understanding what happens when things burn. The discovery that water could be broken down into, or indeed synthesized from, two simpler elementary substances started a chemical revolution in France. The fruits of this revolution are embodied in the very names we now use for these two components, hydrogen and oxygen. However, the path to enlightenment was tortuous, lasting over 200 years. At its peak at the end of the eighteenth century, chemists fell into two distinct camps—those for the new French chemistry, and those against it. Several different names were given to the gases before 'hydrogen' and 'oxygen' triumphed. As it turns out, one of these names is still based on an incorrect theory, and it might have been more appropriate if the names hydrogen and oxygen had been swapped around.

Four Elements, Three Principles

From the sixth century BC, the ancient Greek philosopher Thales taught that water was the primary matter from which all other substances were formed. Perhaps this idea came from water's ready ability to form solid ice, 'earth', or vapours and mists, 'airs'. Other philosophers thought the primary substance was air; others still, fire. It was less common for earth to be thought of in this way, possibly, as Aristotle later wrote, because it was too coarse-grained to make up these fluids. In the fifth century BC Empedokles brought the four 'elements' together—earth, air, fire, and water—and for many centuries it was thought that these made up everything around us. The classic example illustrating this theory was the combustion of wood, as described, for example, in Robert Boyle's classic work from 1661, *The Sceptical Chymist*:

> For if You but consider a piece of green-Wood burning in a Chimney, You will readily discern in the disbanded parts of it the four Elements, of which we teach It and other mixt bodies to be compos'd. The fire discovers it self in the flame by its own light; the smoke by ascending to the top of the chimney, and there readily vanishing into air, like a River losing it self in the Sea, sufficiently manifests to what Element it belongs and gladly returns. The water in its own form boyling and hissing at the ends of the burning Wood betrayes it self to more than one of our senses; and the ashes by their weight, their firiness, and their dryness, put it past doubt that they belong to the Element of Earth.[2]

The four elements were not taken as literally as their names might suggest: Fire also encompassed light, heat, and lightning; Air included vapours such as steam and also smoke; Water meant essentially any liquid, including milk, wine, blood, even acids; and the element Earth referred to most solids including rocks, minerals, and metals.

Paracelsus, the sixteenth-century physician who promoted the use of minerals in medicine and who gave us the word 'zinc', supported a modification suggesting that everything could be broken down (usually with heat) into three principles—the Tria Prima. These were a volatile, fluid species named Mercury; combustible parts, such as oils or fats, generally named Sulphur; and an earthy, solid, involatile component called Salt. These three components should not be confused with the sulfur and mercury that we now know to be chemical elements, or with the sodium chloride salt present in our food. Instead, they represented different qualities or attributes. Common salt itself, a substance of universal familiarity, displays the desired properties of an ideal solid; unlike samples of earth, pure salt is uniform and unchanging in the fiercest of fires, not easily melting or breaking down, and it was these qualities that Salt personified. Mercury, as both planet and metal, was, as we have seen, associated with fast-moving volatility—the god himself darting back and forth between heaven and earth, and the metal easily being boiled to a (very poisonous) vapour. Although in the 1657 work *A Physical Dictionary: or an interpretation of such crabbed words and terms of arts, as are deriv'd from the Greek or Latin, and used in Physick, Anatomy, Chirurgery, and Chymistry*, the entry for Sulphur Philosophorum simply states: 'God knows what the Chymists mean by it', Sulphur represented the very embodiment of flame and, as we have seen, was the ancient mineral known for its flammability.

Given these characteristics associated with the three principles, we see that they are not too distant from the four elements. Referring back to the composition of wood, Paracelsus recognizes the flame-generating component as being the Sulphur, the volatile smoke and water vapour as the Mercury, and the residual ashes as the Salt: 'For that which smokes and evaporates over the fire is

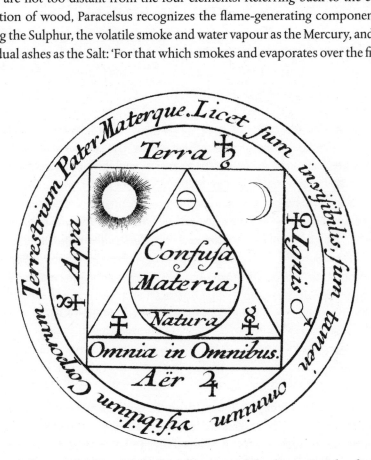

Fig. 30. The interplay between the four elements and the three principles, from an engraving from 1723. The four elements (in Latin *Aqua, Terra, Ignis,* and *Aer*) are arranged around the square, and the symbols of the three principles in the corners of the inner triangle: salt at the top, sulfur bottom left, and mercury bottom right. The symbols for the seven metals are shown: those for the 'imperfect metals' tin, copper, iron, lead, and mercury are arranged around the square (clockwise from the top), and those for the perfect metals, gold and silver, are inside the square in the top left and right corners. The motto around the outside reads in translation: 'Although I am invisible, I am nonetheless the father and mother of all visible earthly bodies.'

Mercury; what flames and is burnt is Sulphur; and all ash is Salt.'[3] The interplay between the four elements and the three principles is shown in an engraving from the 1723 edition of Cornelius Drebbel's tract on the Elements (Figure 30).

Sometimes, particularly in the seventeenth century, two additional principles—Phlegm and Earth—were added to Paracelsus' Tria Prima. These were included to better reflect the different factions obtained after strongly heating organic material derived from plants and animals under the belief that fire could be used to separate matter into its constituent parts. A further modification was suggested by the seventeenth-century chemist Johann Joachim Becher (1635–82), who replaced the Paracelsian principles with three 'earths': vitreous earth (replacing Paracelsus' Salt); inflammable earth (replacing Sulphur) and fluid or mercurial earth (obviously replacing Mercury). According to Becher, these three principles, together with Water, made up all substances; the refinement being necessary since he thought Paracelsus' Tri Prima were actually compounds which could be resolved into his true principles. The important advance made by Becher was that he sought to bring together all the many observations and facts that had been gathered in chemistry, and all the preparations and analyses of substances that had been documented, in order to formulate a coherent theory to explain them. He used the term 'reaction', now so familiar to modern chemists, to explain what happens when two or more species interact, or when one substance is broken down into another. Perhaps most importantly, Becher tried to explain the reactions that take place when substances burn, typified by the 'slow burning' or calcination of a metal to leave a powdery ash.

Phlogiston

Much of Becher's writing is rather difficult to comprehend and it was through the voluminous writings of his disciple, Georg Ernst Stahl (1660–1734) that his ideas, greatly refined and developed, began to disseminate and mature throughout the eighteenth century. Stahl, who for over twenty years taught at the University of Halle in Germany, created a theory of chemistry that prevailed for virtually the whole of the eighteenth century. Central to his doctrine was his take on the principle of flammability: phlogiston.

The name 'phlogiston' itself is from the ancient Greek, meaning 'burnt or inflammable', and is related to the Greek word for 'flame' and more distantly to the English word 'phlegm', via the ancient Greek for 'inflammation'. Stahl states that this is the name he has chosen for this flammable principle, but others used the word in similar ways before him. Becher used it as an adjective, but it also appears as early as 1606, when Nicholaus Hapelius states that it is 'something proper to all sulphur, the essence of that sulphur which is ingrafted in all things'.[4] The Flemish physician Jan Baptist van Helmont suggests that before the Greeks called sulfur divine (thion), it was known as '*Phlogiston*, that is, inflameable: By which Etymology, *Diascorides* soon after said, the best Sulphur was denoted, from its own property, to wit, because it was wholly consumed by the fire.'[5]

Put simply, the theory was this: substances were thought to be flammable because they contained phlogiston. During combustion, the phlogiston left the substance, usually to pass into the air, and what was left, the ash, could not burn since it no longer contained any phlogiston. Certain flammable substances were particularly rich in phlogiston—for example, Stahl thought that soot was the purest form of phlogiston, since it burned without leaving any residue. When things burned, their phlogiston might be lost completely as fire, or just partially, as in the case of burning turpentine, which burns with a very sooty flame—the soot still containing some of the phlogiston.

Even though we now know the phlogiston theory is not correct, it is worth examining more closely. The discoverers of the elements we now call nitrogen and oxygen named their gases using the theory as 'dephlogisticated air' and 'phlogisticated air' respectively. The names we now use emerged as the incorrect phlogiston theory was overthrown—although the word 'oxygen' is still based on an incorrect idea, as we shall see.

Phlogistic Chemistry

Stahl's phlogiston theory was nicely summarized by Richard Watson, the fifth holder of the Chair of Chemistry at the University of Cambridge, who was appointed in 1764 despite the fact that, as he put it, he 'knew nothing at all of Chemistry, had never read a syllable on the subject; nor seen a single experiment in it'.[6] Still, he studied hard and later published a very popular set of chemical essays which included one titled 'Of Fire, Sulphur, and Phlogiston'. After speaking first of elementary fire, he then refers to the fire that enters into the composition of matter—phlogiston.

> Notwithstanding all that perhaps can be said upon the subject, I am sensible the reader will be still ready to ask—*what is phlogiston?* You do not surely expect that chemistry should be able to present you with a handful of phlogiston, separated

from an inflammable body; you may just as reasonably demand a handful of magnetism, gravity, or electricity to be extracted from a magnetic, weighty, or electric body. There are powers in nature which cannot otherwise become the objects of sense, than by the effects they produce; and of this kind is phlogiston.[7]

Watson gives some examples of chemical reactions to illustrate the nature of phlogiston. The first is the burning of a piece of sulfur. We now understand that the sulfur combines with the oxygen from the air to form sulfur dioxide, a choking gas which dissolves in water to give an acidic solution. The view of the phlogistians had nothing to do with the air; instead, it was thought that as the sulfur broke down, the heat locked up in it (which must therefore have been one of its components) and the acid gas (which must be the other component) were liberated.

Watson gives further examples: charcoal gives no vapour when burnt, just a small amount of ash. We would now say that the carbon in the charcoal is combining with the oxygen in the air to give the invisible gas carbon dioxide, and leaving a small amount of inorganic impurities as the ash. The phlogistians thought the charcoal was largely composed of phlogiston, which escaped when it burned, together with a small amount of 'earth', the ash. In contrast, pure alcohol burns completely in air to give carbon dioxide gas and water vapour. The carbon dioxide escaped unobserved, but since the water vapour could be condensed back to liquid water, the followers of the phlogiston theory took alcohol to be composed of water combined with phlogiston.

The final example Watson gives is an extremely important one: the relationship between metals and their ores. Some metals react vigorously when heated in air—magnesium will be familiar today, burning with a dazzling white flame. Magnesium was unknown during Stahl's lifetime and when Watson was writing, but zinc metal was known, and this too burns brilliantly. Other less reactive metals, such as iron, generally react more slowly. We now know that the reaction taking place is the metal combining with the oxygen from the air to form an oxide of the metal, which takes the form of a powdery ash. The view of the phlogistians is that the metal is made from ash and phlogiston, since when it burns (or, in their view, decomposes), the phlogiston escapes and leaves the ash behind.

One of the key pieces of evidence for this view is that the metal may be reformed by returning the phlogiston to the ash. This can be done by heating the ash with a substance rich in phlogiston, such as charcoal. Of course, we now understand that the role of the charcoal, an impure form of carbon, is to remove the oxygen from the metal ash, forming the gas carbon dioxide, and leaving the pure metal element. The phlogistians thought that the metal was being reformed from its components: the ash and the phlogiston from the charcoal.

The phlogiston theory represented an important step, in that that it provided a unifying theoretical basis to much of chemistry. Nevertheless, it was wrong. The error arose partly from the idea that fire broke substances down into their simpler components, but also from neglecting the important role of the air.

Just as water was long thought of as being an element, so too was the air. The realization that air is actually made up of a number of different gases, each with their own unique properties, took many centuries to form. A significant step was made in the seventeenth century with the recognition of 'different types of airs' and how they could be prepared. By the end of the following century, so many new gases had been discovered that a Medical Pneumatic Institution was established in Bristol in order to study their medicinal effects on patients. One of the gases discovered in the eighteenth century was even thought by some to be pure phlogiston itself.

Different Types of Air

Perhaps the first people to recognize the existence of different gases were miners, who would occasionally encounter noxious fluids trapped in pockets underground. Such gases were called damps, originating from the German or Saxon for a vapour or exhalation—presumably a breath from the very bowels of the earth. The number of different damps recognized varied, but could broadly be split into two types: fire-damp and choke-damp.

Fire-damp, also sometimes called fulminating-damp, was most likely the gas we now call methane. Being lighter than air, this would accumulate in the upper parts of the subterranean caverns. While the pure gas would be suffocating, it isn't toxic, so when mixed with air it could still be breathed in and often did not trouble the miners too much. But if the mixture came into contact with a flame or spark, the consequences would be disastrous, since methane forms explosive mixtures with air.

Choke-damp—what we would now recognize as carbon dioxide—presented a completely different hazard. Being heavier than air, it was commonly found at the bottom of pits and mines, and would extinguish flames and suffocate animals. A sudden release of large quantities of the gas would almost certainly result in instant death for the miners. Despite the dangers, the properties of choke-damp were demonstrated in a macabre tourist attraction—the *grotto del cane*, just outside Naples, named after its grisly effect on unsuspecting four-legged victims. Tiberius Cavallo (1749–1809) gives a description of this cave in his *Treatise on the Nature and Properties of Air and Other Permanently Elastic Fluids* from 1781:

> This grotto is about fourteen feet long, and near seven feet high at the entrance. On the floor of it, there is always a stratum of that elastic fluid, which constitutes the choke-damp. It is continually emitted from the earth, through the fissures that may be seen on the ground. The experiments usually shewn to the curious, who visit this grotto, are, first, that of bringing a lighted candle or piece of paper near the floor, which is put out as soon as it comes within about 14 inches of the ground; and, secondly, that of keeping a dog with its head near the ground, for about a minute, so as to oblige him to breathe the noxious fluid, which will soon affect his respiration, deprive him of his strength, and would soon kill him, if he was not immediately brought out into the open air; where, if he is not too far gone, he will gradually recover strength and freedom of respiration.[8]

Later this gas came to be known as 'fixed air', since it seemed to be trapped or 'fixed' in certain rocks, waiting to be liberated using a strong heat. We would now call such rocks carbonates, with chalk, limestone, and marble all being different forms of calcium carbonate.

Chaotic Ghosts

Despite the demonstrations of the properties of carbon dioxide gas, its nature was far from understood. As Cavallo stated: 'The idea, however, that people in general had of this elastic fluid was very confused; the more common opinion, before the time of Mr BOYLE, was, that fixed air, or rather that its effects were owing to a vapour or spirit diffused through the air of some particular places; hence they called it *geist*, i.e. spirit; from whence the word *gas* has been derived.'[9]

This supernatural origin of the word 'gas' is reinforced by the French chemist Antoine-Laurent Lavoisier. In his first published book (1774), he wrote:

> GAS is derived from the Dutch word Ghoast, which signifies Spirit. The English express the same idea by the word Ghost, and the Germans by the word Geist,

which is pronounced Gaistre. These words have too much affinity with that of Gas, to leave any doubt of their derivation.[10]

It's interesting that Lavoisier should speculate in this way on the derivation of the word 'gas', since he notes it was first used and defined by Jan Baptist van Helmont. Born in 1579, van Helmont built on the ideas of Paracelsus and helped to cement the position of chemistry as a true science distinct from alchemy. He recognized that in addition to the vapours that arise on heating liquids such as water, there are other more permanent aerial vapours which do not readily condense into liquids. He writes:

> ...therefore by the Licence of a Paradox, for want of a name, I have called that vapour, Gas, being not far severed from the Chaos of the Auntients. In the mean time, it is sufficient for me to know, that Gas, is a far more subtile or fine thing than a vapour, mist, or distilled Oylinesses, although as yet, it be many times thicker than Air.[11]

This derivation of the word 'gas' from 'chaos' makes sense when we recall the Dutch fricative 'g' is pronounced closer to the 'ch' in 'loch' than to our harder 'g' as in 'get'. In addition to 'gas', van Helmont also invented the word 'blas'. This term never caught on because the idea it encompassed became utterly obsolete: it signified the principle of movement of the stars which influenced various phenomena on Earth, such as the winds and the seasons. As Irish physician Stephen Dickson wrote in his *Essay on Chemical Nomenclature* in 1796, 'Of the latter name, since it has had the ill luck of being entirely overlooked, we need say nothing.'[12]

Dickson didn't like the word 'gas' either, though. In addition to the ghost theory for its derivation, he notes that others suggest it is derived from the German *'gascht'*, meaning 'a frothy ebullition'. He prefers the term 'air', writing: 'Thus, in whatever point of view we consider this *offspring of Chaos*, we find it ill qualified to supersede the term air.'[13]

Van Helmont has been described as the founder of pneumatic chemistry, not because he invented the word 'gas', but because he was perhaps the first to note the formation of a number of different kinds of gases. He noticed the reddish-brown gas (nitrogen dioxide) formed when metals such as silver react with aqua fortis (nitric acid), and also the choking gas formed from burning sulfur (sulfur dioxide). His first-hand experience with the fumes given off by burning coals (the poisonous gas carbon monoxide) very nearly killed him. He writes:

In the year 1643, the day before the Calends of the 11th month called *January*,
I sate beginning to write in a close Chamber; but the cold was great, and I bad
an earthen Pot or Pan to be brought, with burning Coals, that I might sometimes comfort the cold stiffness of my fingers. My little Daughter comes unto
me, who as soon as she sented the hurt or offence, withdrew the Earthen Pan,
and unless she had chanced to come, I being choaked, had perished: For I presently felt about the mouth of my stomach, a sore-threatned swooning; I arose
from my Study; while I would go forth abroad, I fell like a straight staffe, and
was brought away for dead.[14]

This 'treacherous Gas of Charcole' is one of the gases mentioned by English
physician George Thomson, a zealous follower of van Helmont.[15] Writing in
1675, Thomson defines 'Gas' as 'a wild invisible Spirit, not to be imprisoned or
pent up, without damage of what conteins it, arising from the Fermentation of
the Concourse of some Bodies, as it were eructating [belching forth] or rasp-
ing this untamable matter'.[16] In the main text he also refers to 'the Acid Gas of
Sulphur'[17] and speaks of a substance that can 'tame the Gas Sylvestre'.[18] These
are clear references to distinct gases.

The Gas Sylvestre is often mentioned in the original Latin editions of van
Helmont's work, and it has been argued that Paracelsus inspired the use of the
term. In 1566 appeared (posthumously) the first edition of a curious little book
by Paracelsus, *Ex libro de nymphis, sylvanis, pygmaeis, salamandris, et gigantibus*.
This was translated into English in 1941 as *A Book on Nymphs, Sylphs, Pygmies,
and Salamanders, and on Other Spirits*. In this work Paracelsus describes these
fabulous creatures, each of which lives in one of the four elements: 'Those
in the water are nymphs, those in the air are sylphs, those in the earth are
pygmies, those in the fire salamanders.'[19] He says these are not good names,
but he uses them anyway and adds, 'The name of the water people is also
undina, and of the air people sylvestres, and of the mountain people gnomi,
and of the fire people vulcani rather than salamandri.' In *A Chymical Dictionary
Explaining Hard Places and Words met withal in the Writings of Paracelsus, and other
Obscure Authours*, published in 1650, the author states, '*Sylvestres*, or *sylvani* are
aiery men, and aiery spirits, sometimes they are taken for woodmen that are
strong giganticall men.'[20]

Paracelsus says that each of the four spirits has their own Chaos, by which
he means their own space or habitat. Just as people have their abode in air,
and fishes in water, 'the undinae have their abode in water, and water is given
to them as to us the air, and just as we are astonished that they should live in
water, they are astonished about our being in air'.[21] He later gives the useful

advice that 'One who has a nymph for a wife, should not let her get close to any water, or at least should not offend here while they are on water,' for they are able to easily escape into their element.[22] With regards to habitat: 'The same applies to the gnomi in the mountains: the earth is their air and is their chaos.' These gnomes (the mining demons from Chapter 2) do not just live in caverns so often found underground; 'they walk through solid walls, through rocks and stones, like a spirit; this is why these things are all mere chaos to them, that is, nothing'. He adds that 'the sylvestres are closest to us, for they too maintain themselves in our air'.[23]

Whether or not he was inspired by Paracelsus' writings of the Sylvestres whose Chaos was the air, van Helmont certainly used the term 'Gas Sylvestre' in a different way, usually to denote what we now know as the gas carbon dioxide. In the English edition of his collected works, this is translated as 'wild spirit': we have the 'wild spirit belched forth' when an acid such as vinegar dissolves shells. More dramatic is the preparation of carbon dioxide gas in a sealed bottle. The carbonate and acid are put into a strong bottle, then: 'straightway let the neck of the Glasse be shut by melting it, which is called *Hermes Seal*: As soon as the voluntary action shall begin, and the Vessel is filled with a plentifull exhalation (yet an invisible one) and however it may be feigned to be stronger than Iron, yet it straightway dangerously leapeth asunder into broken pieces.'[24]

This effervescence from acid and carbonates had been known for a long time (it might even be referred to in the Old Testament, as we shall see) but the relationship between carbonates, acid, and gas were not properly investigated until the Scottish physician and chemist Joseph Black (1728–99) turned his attention to the matter in the middle of the eighteenth century while trying to find an effective treatment for bladder stones.

Fixed Air

Black's work, published in 1756, focuses on what he termed '*magnesia alba*', which we would now call magnesium carbonate; it also looks at 'calcerous earths' (calcium carbonate) and 'alkalis' (the water-soluble potassium carbonate, sodium carbonate, and ammonium carbonate). He notes that, when treated with various acids, 'Magnesia is quickly dissolved with violent effervescence, or explosion of air...'[25] He also records a decrease in mass when the magnesia is heated, and notes that after heating, the residue—calcined magnesia—no longer effervesces. This, he correctly interprets, is because strong heat drives the gas from the mineral: 'We may therefore safely conclude, that the volatile matter, lost in the calcination of *magnesia*, is mostly air; and hence the calcined *magnesia* does not emit air, or make an effervescence, when mixed with acids.'[26]

Initially Black just refers to the gas (carbon dioxide) as 'air', but when locked up or 'fixed' in a variety of (carbonate) minerals, he calls it *fixed air*. 'It is sufficiently clear, that the calcareous earths in their native state, and that the alkalis and magnesia in their ordinary condition, contain a large quantity of fixed air, and this air certainly adheres to them with considerable force, since a strong fire is necessary to separate it from magnesia...'[27] Years later, perhaps after it was found that other gases may also be liberated from solids and would equally be deserving of the term 'fixed air', Black seems to prefer the name 'mephitic air' for his gas. This term had historically been used for more offensive-smelling odours.

Black demonstrated the physical and chemical properties of the gas we now know as carbon dioxide; he showed it to be different from normal air, and understood how it could be prepared from different carbonates either by heating or by the action of acids. However, the elegant line of reasoning he presented may actually have helped to prolong the misunderstanding of the production of another key gas, the one we now know as hydrogen—a gas also previously observed by van Helmont.

Carbon dioxide might be the oldest gas observed to form, but it is not an element. Hydrogen gas is an element, but both gases were initially thought of as modifications of air. There was no realization that one was a compound and the other an element until the end of the eighteenth century. The first element in the periodic table could not get its final name until the different gases in the atmosphere were recognized and their chemical properties understood.

Early Accounts of Hydrogen

Hydrogen was first described by van Helmont in the first half of the seventeenth century, many years before it was collected and carefully studied. He observed that while distilling nitric acid (aqua fortis) alone, nothing untoward occurs, but 'if a dissolvable mettal be added unto it, it brings forth a Gas, so as that if the glass be well stopt with morter, although most strong, it breaks in pieces'.[28]

Van Helmont does not seem to have discovered the flammability of the hydrogen gas he produced, but it was described by Robert Boyle in his *Tracts...Containing New Experiments, Touching the Relation betwixt Flame and Air* from 1672. Here he describes an experiment in which acid is poured onto filings of steel (freshly made, and not those 'commonly sold in shops to Chymists and Apothecaries', which are usually rusty):

> the mixture grew very hot, and belch'd up copious and stinking fumes…whencesoever this stinking smoak proceeded, so inflammable it was, that upon approach of a lighted candle to it, it would readily enough take fire, and burn with a blewish and somewhat greenish flame at the mouth of the viol for a good while together; and that, though with little light, yet with more strength than one would easily suspect.[29]

Over sixty years later, in 1736, John Maud demonstrated before members of the Royal Society the flammability of hydrogen collected in bladders—the eighteenth-century equivalent of modern latex balloons. He compared the flame produced with that of samples of 'fire-damp' collected by a colleague from a coal mine. In the report of the demonstration, he writes: 'It is very well known to every one versed in Chemical Affairs, that most metals emit great Quantities of sulphureous Vapours, during the Effervescence which they undergo in their Solutions in their respective *Menstrua*, or *Solvents*. Of these Fumes Iron emits a great Quantity whilst it is dissolving in Oil of Vitriol, which are very inflammable, and not easily to be condens'd.'[30]

It was noted that the flame of the burning 'sulphureous vapours' (hydrogen) was somewhat different in colour to that of the burning sample of gas collected from the mine. We now know that the yellow colour of a flame is due to tiny particles of glowing carbon (soot) that are formed when hydrocarbons such as natural gas, petrol, or candle wax burn without enough oxygen. Having no carbon in its composition, hydrogen burns with a blue flame which can be almost invisible. A few decades later, this difference was one of the key means of distinguishing hydrogen from methane, the other common flammable gas, but at the period of these observations, it was thought that all flammable gases were the same.

We now understand that the hydrogen evolved comes from the aqueous acid during its reaction with the metal, but at the time, it was thought that this 'sulphureous vapour' originated from the metal itself. Maud writes, 'What is worthy of Notice in this Experiment is, that all the Air which fill'd the Bladders was as it were generated *de novo* out of the Mixture, or else recover'd from being lock'd up in the Body of the Metal in an unelastic State.'

Even though Maud clearly prepared and isolated hydrogen gas, and studied its combustion (one bladder 'went off like a Gun, with a great Explosion') he is not generally credited with being its discoverer. This honour is usually bestowed on Henry Cavendish (1731–1810), one of the most brilliant but socially awkward scientists of the eighteenth century.

Cavendish's Inflammable Air

Although he did not begin life with great wealth, Henry Cavendish became one of the richest men in England after the deaths of various noble relatives—his father being the son of the Duke of Devonshire, and his mother the daughter of the Duke of Kent. Despite his vast inherited wealth, he lived frugally as a recluse and wore clothing from a bygone era. He was incredibly shy, fleeing from the sight of women (his servants were threatened with dismissal if he saw them), and he tolerated only the company of his fellow scientists at the Royal Society. Hating confrontation, he published only a fraction of his work. Had the world seen his unpublished research, science would have been significantly advanced.

Cavendish's first published work was titled *Three Papers, Containing Experiments on Factitious Air*. It appeared in 1766, thirty years after Maud's observations on 'inflammable air'. Cavendish begins by defining 'factitious' and 'fixed' air:

> By factitious air, I mean in general any kind of air which is contained in other bodies in an unelastic state, and is produced from thence by art.
>
> By fixed air, I mean that particular species of factitious air, which is separated from alkaline substances by solution in acids or by calcination; and to which Dr Black has given that name in his treatise on quicklime.[31]

Just as Black made meticulous measurements concerning the formation of fixed air (carbon dioxide) from the action of heat or acid on various carbonate minerals, so did Cavendish for the formation of inflammable air (hydrogen) from the action of acid on the metals zinc, tin, and iron. He collected the gas in glass vessels by the displacement of water, and in bladders which enabled him to determine its density. As well as simply burning the gas, he also mixed it with different proportions of air and noted the varying degrees of the explosions on applying a light. Crucially, he noticed that for each metal, as long as the same mass of the particular metal was used, it always produced the same quantity of gas, regardless of the acid used. Cavendish thought the liberated gas was the phlogiston initially contained in the metal:

It seems likely from hence, that, when either of the above-mentioned metallic substances are dissolved in spirit of salt [hydrochloric acid], or the diluted vitriolic acid [sulfuric acid], their phlogiston flies off, without having its nature changed by the acid, and forms inflammable air.[32]

With hindsight, it is easy for us to fault the idea that phlogiston is released from metals when they dissolve in acid, but actually, this was a logical extension of Black's work on carbonates. Black had correctly shown that if certain metal carbonates are heated, the fixed air they contain is driven out, and a metal ash or calx is formed. Adding acid to the carbonate also drives out their fixed air. Along with other phlogistians, Cavendish believed that when metals burned, they released their phlogiston to leave an ash. He now found that adding acids to metals similarly drove out their phlogiston, the inflammable air which he now captured.

This identification of inflammable air (hydrogen) with phlogiston was further reinforced when it was later discovered that certain metals can be reformed by heating their ashes in an atmosphere of this gas; this reaction was seen as reforming the metal from its component ash and phlogiston. This experiment was first performed by Joseph Priestley in 1782. Priestley was a staunch believer in phlogiston, still clinging to the theory after all other scientists had abandoned it and taking it to his deathbed in 1804. And this despite the fact that he discovered the key piece in the puzzle: the gas we now call oxygen.

Drebbel's Submarine

Just as hydrogen gas had been prepared in the seventeenth century before it was systematically studied in the eighteenth, oxygen was also prepared prior to its official discovery by Joseph Priestley in England, and independently by Carl Wilhelm Scheele in Sweden. One of the most intriguing theories is that it was prepared by the maverick Dutch inventor Cornelius Drebbel (1572–1633) and utilized by him in his submarine, demonstrated in the Thames in London before King James I in the early seventeenth century. Boyle's account states that Drebbel had some sort of liquid with which he was able to refresh the air in the enclosed space of the submarine after the twelve operators and passengers had been breathing it for some time:

For when from time to time he perceiv'd, that the finer and purer part of the Air was consum'd, or over clogg'd by the respiration, and steames of those that went in his ship, he would, by unstopping a vessel full of this liquor, speedily restore to the troubled Air such a proportion of Vitall parts, as would make it againe, for a good while, fit for Respiration, whether by dissipating, or precipitating the grosser Exhalations, or by some other intelligible way, I must not now stay to examine.[33]

Fig. 31. A tantalizing engraving from the Dutch translation of Cornelius Drebbel's book on the elements from 1702. The bubbles evolved would be oxygen gas if the flask being heated contains saltpetre (potassium nitrate).

Passages in Drebbel's book *Treatise on the Nature of the Elements*, first published in German in 1608, mention the effects of heat on saltpetre (potassium nitrate). This is a substance that we now know decomposes with a moderate heat to give out oxygen gas. There is even a suggestive picture in some editions of Drebbel's book (Figure 31) that shows a glass retort being heated over a fire with its neck immersed in water and bubbles coming out. Some have proposed that Drebbel collected the gas and used this to revive the air in his submarine. However, what makes the air in a confined space unpleasant to breathe is not the lack of oxygen but the build-up of the carbon dioxide being exhaled. This is easily removed by a solution of strong alkali, and this is what is formed on dissolving in water the solid residue left after strongly heating saltpetre.

While it is pure speculation whether or not Drebbel actually collected and used oxygen gas, it is certainly true that this element was prepared and collected in the 1720s by Stephen Hales (1677–1761). Under the section 'Analysis of the Air' in his book *Vegetable Staticks*, published in 1727, Hales describes heating pretty much anything he could lay his hands on—peas, a fallow deer's horn, amber, and hog's blood—in order to measure, by the displacement of water, how much 'air' it contained. He reports that from 'half a cubick inch of *Nitre*, mixed with the calx of bones, there arose 90 cubick inches of air'.[34] This would have been oxygen gas. Sadly, Hales did not study the properties of any of the gases he prepared, thinking they were simply modified forms of air. If he had done so, science might have been advanced by some fifty years.

The first person who did prepare and study the properties of oxygen was the Swedish apothecary Carl Wilhelm Scheele (1742–86), although the accounts of his discovery were not published until 1777, two years after those from Joseph Priestley appeared in England.

Scheele's Fire Air

At the age of fourteen, Carl Wilhelm Scheele became the apprentice of an apothecary and so gained access to the scientific apparatus, chemicals, and textbooks that would help him not only to learn his trade, but also to develop into one of the most skilful and respected chemists of the eighteenth century. Despite limited means and resources, he made a large number of important discoveries in chemistry, including the first preparation of chlorine gas and the identification of several new metallic elements. In the late 1760s and early 1770s, Scheele was trying to understand the nature of fire, 'since it is impossible to make experiments without heat and fire'. But before long, his investigations were heading in a different direction. As he describes in his book: 'I soon found, that without knowing the *Air*, it is impossible to form a true judgment on the phenomena of Fire. After a series of experiments I observed, that *Air really makes part of the compound of Fire, and is a constituent part of the flame and of sparks.*'[35]

Unlike his forerunner Hales, Scheele realized that the gases given out from different reactions are not just modifications of atmospheric air, but may be 'distinct varieties of air'. He soon discovered that atmospheric air 'must be composed of Elastic Fluids of Two Kinds'. Scheele was able to remove the oxygen from a sample of air, for example, by burning sulfur in it, and then removing the sulfur dioxide formed by absorbing it in alkali. This left almost pure nitrogen gas, which he distinguished from normal air and the other known gases, hydrogen and carbon dioxide: unlike normal air, it did not support combustion; it was not flammable like hydrogen, and while it easily extinguished flame like carbon dioxide, unlike this latter species, his new gas did not react with alkali.

The other component in air (oxygen) he prepared by a variety of methods, including heating potassium nitrate, just as Hales had done before; only Scheele realized he had isolated something quite new. 'I filled a ten-ounce glass with this air and put a small burning candle into it; when immediately the candle burnt with a large flame, of so vivid a light that it dazzled the eyes.'[36]

Scheele then describes how, by mixing this oxygen with his previously isolated nitrogen, normal air might be recreated: 'I mixed one part of this air with three parts of air, wherein fire would not burn; and this mixture afforded air, in every respect similar to the common sort.' Scheele also named his new gases, but not with the names we know today, oxygen and nitrogen. He writes: 'Since this air is absolutely necessary for the generation of fire, and makes about one-third of our common air, I shall henceforth, for shortness sake call it *empyreal air* (literally *fire-air*) the air which is unserviceable for the fiery phenomenon, and which makes about two-thirds of common air, I shall for the future call *foul air* (literally *corrupted air*).'[37]

The quotations from Scheele's work are from the 1780 English edition of his work, translated by John Forster. While Forster uses the term 'empyreal air' throughout the book, he does note in brackets the more literal term 'fire-air'. Similarly, while using 'foul air' for nitrogen, he also mentions an alternative translation, 'corrupted air'. Scheele himself complained that Forster's translation of his book was inaccurate, and in 1931 a more faithful translation was prepared by Leonard Dobbin, who used the terms 'Fire Air' for the German original 'Feuer Luft' and 'Vitiated Air' for 'Verdorbene Luft'. The Oxford English Dictionary defines 'vitiated' as 'corrupted, spoiled, impaired'; Scheele states that 'Verdorbene Luft' was a term already known. The term 'vitiated air' had also been used before in English works in the first half of the eighteenth century, usually just meaning 'air that would not sustain life or allow things to burn'.

Scheele's careful determinations of the densities of his new gases added further support to the fact that common air is just a mixture of the two. He correctly found his foul air (nitrogen) to be slightly lighter than atmospheric air, and his empyreal air or fire air (oxygen) to be slightly heavier.

After detailing the preparations of his fire air, Scheele describes a number of experiments with it, including burning various substances in it, such as coal, sulfur, and phosphorus. He even filled a bladder and tried breathing his new air himself (Figure 32). He writes: 'I tied up the bladder, removed it from the retort, and fixed a tube to its orifice; and having quite emptied my lungs from Air, I began to breath the Air out of the bladder. This succeeded so well, that I was able to take *forty* inhalations before it became troublesome to me.'[38]

The air from the bladder became uncomfortable to breathe because of the accumulation of carbon dioxide, even though there was still oxygen left in it. Scheele discovered how he could continue breathing the air longer (sixty-five inhalations) if he added some alkali. This alkali would have absorbed the carbon dioxide he was breathing out, and his success supports the idea that this may have been the method used by Drebbel to improve the air in his submarine.

Fig. 32. An eighteenth-century engraving showing how to breathe gas from a bladder.

Scheele interpreted his findings using the phlogiston theory. Essentially, he believed the two gases that make up atmospheric air differed in how they interacted with phlogiston—the major component (nitrogen) 'attracts not the phlogiston' while the minor component (oxygen) does attract it, and it is this attraction to phlogiston that enables things to burn. Further, he actually thought that fire was made from the combination of phlogiston with his new 'Fire Air'—hence the name.

Scheele was not the only person to identify nitrogen gas—like oxygen, this gas was also independently discovered by Joseph Priestley around the same time, and perhaps earliest of all by Henry Cavendish, who typically did not bother to publish this major discovery but did clearly communicate his findings to Priestley. However, it was also independently discovered in 1772 in Scotland by one of Joseph Black's pupils, Daniel Rutherford.

Foul, Vitiated, Mephitic Air

It is curious that despite being by far the most abundant element from Group 15 in the periodic table of the elements, nitrogen was the last to be discovered—phosphorus being discovered in the late seventeenth century, and, arsenic, antimony, and bismuth all being known much earlier. This highlights the difficulties in understanding the true nature of the air. When recognizing the 'official' discovery of this gas, the important thing is that it should be distinguished from the other known gases, and particularly from that which supports neither life nor combustion—Black's fixed air, carbon dioxide. The first clear publication of these distinctions appears in the doctoral dissertation of Daniel Rutherford (1749–1819), who later became a professor of botany at the University of Edinburgh and invented the maximum and minimum thermometer still used today. Black assigned to Rutherford the task of investigating the air that remained after various carbon-containing substances had been burned in air and the fixed air so formed had been dissolved in alkali.

Rutherford starts off by looking at carbon dioxide, which, as most likely directed by his master Black, he also calls 'mephitic air': 'By Mephitic Air, which some call Fixed Air, I understand, with the distinguished Prof. Black, that singular species of air which is fatal to animals, which extinguishes fire and flame, and which is attracted with great avidity by quick-lime and alkaline salts.'[39]

He mentions how it can be formed in the bowels of the earth, and in the *grotto del cane* we met earlier. He also appreciates how it arises from the lungs of animals through respiration, also from combustion, and finally from chemical processes such as the action of acids on carbonate minerals. But the test for this gas which readily distinguishes it from other 'airs', a reaction known by many young students, is that it has a strong affinity for lime-water (a solution of calcium hydroxide), with which it gives a milky-white precipitate. Rutherford has a clear understanding of carbon dioxide, but he also thinks something else can change the air to make it no longer support life. He writes: 'But, by the respiration of animals, wholesome and good air not only becomes in part mephitic, but it also suffers another singular change. For, after all the mephitic air has been separated and removed from it by means of caustic lye, still what remains does not become in any way more wholesome; for although it produces no precipitate in lime-water, it extinguishes both flame and life no less than before.'[40]

By the addition of caustic lye (potassium hydroxide solution), any carbon dioxide present is completely removed. If all the oxygen initially present in a sample of atmospheric air had been converted to carbon dioxide through respiration or combustion, once removed, all that remains is essentially the nitrogen. This was the method used by Scheele, and indeed by Cavendish before. The same gas could also be produced by burning other species in air, such as sulfur and phosphorus, and again using alkali to remove their acidic products of combustion. Rutherford concluded that in each case the air is modified by the addition of the phlogiston from those flammable substances: 'From these things we may also conclude that malignant air is composed of atmospheric air united with phlogiston and, as it were, saturated.'[41] Rutherford did not name this phlogiston-saturated air, but Joseph Priestley, who independently reached the same conclusion as Rutherford, did give it a name: 'phlogisticated air'.

Priestley's Different Kinds of Air

The dissenting clergyman Joseph Priestley (1733–1804) was one of the more colourful characters in the history of chemistry. In stark contrast to Cavendish, Priestley was never one to shy away from controversy—he eventually left England for America in 1791 after a mob burned down his Birmingham home and the church where he was minister in the so-called Priestley Riots.

Priestley was a prolific author; when asked exactly how many books he had written, he is reported to have answered, 'Many more, Sir, than I should like to read'. His earliest scientific publications were concerned with discoveries in electricity and light, but he later shifted his attention to the study of gases after moving next door to a brewery which produced large quantities of carbon dioxide. In 1772 he published *Directions for Impregnating Water with Fixed Air*, which detailed the production of carbonated water, or soda water, which he is credited with inventing. Two years later saw the first volume of his *Experiments and Observations on Different Kinds of Air*. In this work, he describes the preparation of a number of gases including fixed air (carbon dioxide), inflammable air (usually hydrogen), marine acid air (hydrogen chloride), and alkaline air (ammonia). The last two, being extremely soluble in water, he ingeniously collected by the displacement of mercury.

Priestley, like Cavendish, Scheele, and Rutherford, also found that part of the air is used up during the combustion of different species, and if the products are absorbed, the resulting air (essentially just nitrogen gas), now diminished in volume, no longer supports life or further combustion. Just as Rutherford had done, Priestley interpreted these findings as being due to the transfer of phlogiston to the air, and on this basis, he suggested his name for the resultant gas: 'On this account, if it was thought convenient to introduce a new term (or rather make a new application of a term already in use among chymists) it might not be amiss to call air that has been diminished, and made noxious by any of the processes above mentioned, or others similar to them, by the common appellation of *phlogisticated air*.'[42]

Unlike Scheele, whose findings were delayed by almost two years before being announced to the world, Priestley liked to publish his discoveries as soon as possible—often in a form lacking the elegant train of thought so clearly evident throughout Scheele's work. When Priestley talked of his isolation of nitrogen, he had not yet found oxygen. This latter discovery, which he says took place on 1 August 1774, was first announced in the Philosophical Transactions of the Royal Society in 1775 and later that year in the second volume of his *Experiments and Observations on Different Kinds of Air*. After describing some of the gases given out on heating various substances using rays from the Sun focused through a lens, he adds: 'But the most remarkable of all the kinds of air that I have produced by this process is, one that is five or six times better than common air, for the purpose of respiration, inflammation, and, I believe, every other use of common atmospherical air. As I think I have sufficiently proved, that the fitness of air for respiration depends upon its capacity to receive the *phlogiston* exhaled from the lungs, this species may not improperly be called, *dephlogisticated air*.'[43]

Since Priestley believed substances burned by transferring their phlogiston to the air, unlike his phlogisticated air (nitrogen), which was saturated with phlogiston and so could receive no more and therefore only extinguish flame, this new gas (oxygen) was devoid of the mythical substance, so it could support combustion much better than normal air by taking in the phlogiston from the burning matter. 'A candle burned in this air with an amazing strength of flame; and a bit of red hot wood crackled and burned with a prodigious rapidity, exhibiting an appearance something like that of iron glowing with a white heat, and throwing out sparks in all directions. But to complete the proof of the superior quality of this air, I introduced a mouse into it; and in a quantity in which, had it been common air, it would have died in about a quarter of an hour, it lived, at two different times, a whole hour, and was taken out quite vigorous…'[44]

In his book, he foresees that this new gas may be of use in medicine but urges caution for use by healthy individuals: 'though pure dephlogisticated air might be very useful as a *medicine*, it might not be so proper for us in the usual healthy state of the body: for, as a candle burns out much faster in dephlogisticated than in common air, so we might, as may be said, *live out too fast*, and the animal powers be too soon exhausted in this pure kind of air. A moralist, at least, may say, that the air which nature has provided for us is as good as we deserve.'[45]

These thoughts didn't stop Priestley trying to breathe pure oxygen gas himself, just like Scheele before him. He writes: 'The feeling of it to my lungs was not sensibly different from that of common air; but I fancied that my breast felt peculiarly light and easy for some time afterwards. Who can tell but that, in time, this pure air may become a fashionable article in luxury. Hitherto only two mice and myself have had the privilege of breathing it.'[46] Of course, he did not know at this time that Scheele had also done so.

One of the substances Priestley heated in order to obtain his dephlogisticated air was *mercurius calcinatus* per se (mercury oxide). Since he was suspicious of the authenticity of the sample he had, on a visit to Paris in October 1774, 'knowing that there were several very eminent chymists in that place', he sought out another sample. Over dinner, he mentioned to his Parisian hosts 'the surprize at the kind of air which I had got from this preparation'.[47] His host was the final player who is also sometimes described as a discoverer of oxygen, the chemist who was to dispel the phlogiston theory once and for all: Antoine-Laurent Lavoisier.

The Father of Modern Chemistry

In contrast to both Scheele and Priestley, Lavoisier (1743–94) was born into a privileged family and received a thorough education, not just in the sciences but also in finance and law. Now regarded as one of France's most eminent scientists, tragically he was one of the many guillotined during the height of the Revolution, primarily because of his involvement with the organization responsible for collecting taxes. Within two years of his death, Lavoisier was formally pardoned on the realization that far from embezzling from the state and owing money, he had actually done the reverse.

Some of Lavoisier's earliest chemical investigations were concerned with the problem of combustion. In his first book, published in French in 1774 and translated two years later as *Essays Physical and Chemical*, he carefully burned various substances such as sulfur, phosphorus, and a variety of metals in air and noted that part of the air was used up. He confirmed that substances heated in vessels where the air had been evacuated underwent no change, proving the air to be necessary. Most importantly, he devised apparatus that enabled him to weigh not only the substances themselves before and after combustion, but also the air before and after. He noted that what the substance gained in mass, the air lost. While others had noted the changes in mass before, he is regarded as the first to show that the total mass is conserved and just redistributed during such reactions. Clearly this gain in mass of the substance burned did not fit well with Stahl's theory, which supposed that it lost phlogiston in the process.

Lavoisier had clearly shown that part of the air is absorbed when various substances such as metals form their ash or calx. What he really needed to do to complete his understanding was to reverse this process— that is, to reform the metal and gas without the aid of any other substance. He was eventually able to do this in November 1774.

Whether or not Priestley's dinner conversation about the gas he obtained on heating mercury oxide really provided the inspiration for Lavoisier to repeat the experiment is arguable (although not for Priestley). Other French chemists had reported the formation of mercury on heating the oxide, and at around the same time he heard from Priestley, Lavoisier witnessed a demonstration of this process by his colleagues. The first account in English of his preparation of oxygen appears as an appendix added to the *Essays* in 1776, entitled 'On the Nature of the Principle which is combined with Metals during their Calcination, and Occasions an Increase in their Weight'.

In his usual thorough style, Lavoisier first proved that *mercurius calcinatus* per se (mercury oxide) was a true metal calx which behaved like others by forming fixed air (carbon dioxide) and metal on heating with charcoal. He then heated the mercury oxide without any charcoal and collected the evolved gas (oxygen). He tested the properties of this gas to show, much to his surprise, that it was not fixed air: 'far from being fatal, like it, to animals, it seemed, on the contrary, more proper for the purpose of respiration; candles and burning bodies were not only *not* extinguished by it, but burned with an enlarged flame in a very remarkable manner; the light they gave was much greater and clearer than in common air'.[48]

While Lavoisier clearly established that the gas was *not* fixed air, at this point in his original paper (he corrects it in later versions), he does not state that this gas is a new, distinct species. Instead he writes: 'All these circumstances fully convinced me that this air was not only common air, but that it was even more respirable, more combustible, and consequently more pure even than the air in which we live.'

Lavoisier does, however, go on to give the very first correct account of the calcination of metals, and the reverse process, the reduction of metallic calces with charcoal to reform the metals: 'It seems to be proved from hence, that the principle which combines with metals during their calcination, and which occasions the augmentation in their weight, is nothing but an exceedingly pure portion of the air which surrounds us, which we respire, and which passes, in this process, from a state of expansibility to that of solidity.'[49]

Lavoisier does not yet name this new gas—in a few subsequent papers he uses Priestley's term 'dephlogisticated air', and then 'pure air' or 'highly respirable air'. He chooses his name—oxygen—only after further experiments on the gas and the effects it produces.

The Acid Former

The papers announcing Lavoisier's latest findings were translated into English by Thomas Henry and published in 1783 under the title *Essays on the Effects Produced by Various Processes on Atmospheric Air; with a Particular View to and Investigation of the Constitution of the Acids*. Henry, who had previously translated Lavoisier's *Essays*, notes in his preface that Lavoisier had not yet published the promised second volume of his previous work, since 'he appears to have been principally occupied in an attempt to overthrow Stahl's doctrine of phlogiston, and in an investigation of the nature and constitution of the acids'.[50]

The essay which concerns us is Essay VIII, which shows that Lavoisier is interested in the true principles or elements which make up chemical substances. He starts by outlining how the ancient chemists broke things down into their ultimate components by heat, and that, for a while, salt was thought to be a component, but then this was further found to be composed of an acid with a base.

Lavoisier goes on to say that now it is possible to deduce what the acid and base themselves are made of. He states:

> In the foregoing essays I have endeavoured to prove as clearly as is possible by physics and chemistry, that the very pure air which Dr Priestley has denominated *dephlogisticated air*, enters, as a constituent part, into the composition of several acids, and especially into that of the phosphoric, vitriolic, and nitrous acids.
>
>   Many additional experiments enable me to generalise this doctrine, and to declare that this pure and highly respirable air, is the constitutive principle of acidity; that this principle is common to all acids; and that the difference by which they are distinguished from each other is produced by the union of one or more principles besides this air, so as to constitute the particular form under which each acid appears.[51]

There is sense to Lavoisier's reasoning here. Sulfur, phosphorus, and carbon all burn in oxygen to form compounds that give an acidic solution when dissolved in water. Not all elements produce acids when they burn, and the role of the water is also crucial. But for Lavoisier, the oxygen was the most important component of acids, and it therefore suggested the name for the element: 'These facts being, in my opinion, firmly established, I shall in future distinguish dephlogisticated or highly respirable air, in a state of combination or fixity, by the name of the *acidifying* principle, or, if any person prefer to express the same signification by a Greek word, the *oxyginous* principle.'[52]

Note that here Lavoisier is not actually referring to the gas oxygen, but the principle, or element, that may enter into the composition of different substances. He believes oxygen gas is a compound of this principle with fire, heat, or light. He does not yet use the word 'oxygen', instead writing: 'That the acidifying or oxyginous principle, combined with the matter of fire, of heat, and of light, forms pure or dephlogisticated air.'

Despite Lavoisier's persuasive arguments, the translator Henry was not convinced. Even though 'many began even to doubt its existence, and to regard it as a creature of the imagination', Henry still stuck to the phlogiston theory and rejected Lavoisier's correct explanation of combustion. The reason? 'The existence of phlogiston, however, has not only been proved, but Dr Priestley has clearly shewn that phlogiston and inflammable air are the same thing…and that this air is capable of being wholly absorbed in the reduction of metals, and of restoring to the calxes their pristine metallic splendor and malleability.'[53]

Phlogiston Bottled

The experiment that Henry is referring to is Priestley's discovery that metals may be formed from their calxes (oxides) by heating them in hydrogen gas. In the introduction to this paper, published in the same year as Henry's translation of Lavoisier's papers, 1783, Priestley tells how even he had almost been convinced by Lavoisier's arguments:

> Of late it has been the opinion of many celebrated chemists, Mr LAVOISIER among others, that the whole doctrine of phlogiston had been founded on mistake, and that in all cases in which it was thought that bodies parted with the principle of phlogiston, they in fact lost nothing, but on the contrary, acquired something; and in most cases, an addition of some kind of air; that a metal, for instance, was not a combination of two things, *viz.* an earth and phlogiston, but was probably a simple substance in its metallic state; and that the calx is produced not by the loss of phlogiston, or of any thing else, but by the acquisition of air.
>
> The arguments in favour of this opinion, especially those which are drawn from the experiments of Mr LAVOISIER made on mercury, are so specious, that I own I was myself much inclined to adopt it. My fried Mr KIRWAN, indeed, always held that phlogiston was the same thing with inflammable air; and he has sufficiently proved this from many experiments and observations, my own as well as those of others. I did not, however, accede to it till I discovered it by direct experiments, made with general and indeterminate views, in order to ascertain something concerning a subject which had given myself and others so much trouble.[54]

Priestley then describes his experiment, where he uses the sun's rays and a powerful lens to heat the calx of lead (lead oxide) in an atmosphere of inflammable air (what he took to be hydrogen) confined in a glass vessel over water. He was extremely excited and pleased to see lead metal forming, and the water level rise as the gas confined in the vessel was used up. To him, this was convincing proof that the calx was absorbing the inflammable gas to become the metal. This fitted in perfectly with the phlogiston theory: the flammable gas (the hydrogen) must be pure phlogiston, which when added to the calx gave the metal, as predicted following Stahl's doctrine. What Priestley missed was the water that was also formed during the reaction—and he missed it because his experiment was carried out over water. His method of producing his inflammable air was also problematic since, in addition to hydrogen, other flammable gases were present, notably carbon monoxide. He would not appreciate this until a couple of years later, but it didn't affect his conclusions at this point.

Priestley's crucial experiment led Thomas Henry to write with dramatic flair that 'the being of phlogiston can be no longer doubted; as Dr Priestley has literally given *"to airy nothing a local habitation and a name"*; and has embodied and rendered visible this Proteus which has so long eluded the grasp and sight of the chemist'.[55]

After reporting the apparent absorption of phlogiston into the metallic calx, in the second half of his paper, Priestley includes his attempts to reproduce one of the most pivotal experiments in the history of chemistry: the reaction between hydrogen and oxygen gases to form water. This experiment is so important that it actually gave rise to the very name used for the most abundant element in the universe—hydrogen—from Greek roots meaning 'water-former'. Priestley first heard of the experiment from its discoverer: our brilliant and reclusive scientist, Henry Cavendish.

The Synthesis of Water

Cavendish's notebooks show that he first synthesized water from hydrogen and oxygen gases in July 1781 but, characteristically, he did not publish until three years later, even though he openly told others of his experiments, including Priestley and Lavoisier. At this time he was still a phlogistian, and his experiments were made 'principally with a view to find out the cause of the diminution which common air is well known to suffer by all the various ways in which it is phlogisticated, and to discover what becomes of the air thus lost or condensed'.[56] This was indeed a curious problem, since if, during combustion, the air was supposed to be receiving the phlogiston from the burning substance, why did it seem to disappear? Cavendish's attention to the combustion of hydrogen was directed by the slightest of observations recorded by Priestley. He writes:

> In Dr PRIESTLEY's last volume of experiments is related an experiment of Mr WARLTIRE's, in which it is said that, on firing a mixture of common and inflammable air by electricity in a close copper vessel holding about three

pints, a loss of weight was always perceived, on an average about two grains, though the vessel was stopped in such a manner that no air could escape by the explosion. It is also related, that on repeating the experiment in glass vessels, the inside of the glass, though clean and dry before, immediately became dewy; which confirmed an opinion he had long entertained, that common air deposits its moisture by phlogistification.

Cavendish approached the issue with characteristic precision. Unlike Priestley, Cavendish understood that there were different types of flammable gas, and he used pure hydrogen prepared (as he had documented in 1766) from the action of acid on metals such as iron or zinc. When repeating Warltire's experiment many times and on larger scales, the skilled Cavendish never perceived any significant loss of mass. He confirmed the vessel became dewy, but that no soot was deposited (unlike Warltire had observed, since his hydrogen was far from pure). Cavendish precisely determined the maximum volume of hydrogen that would react with the oxygen part of the air and leave behind the unchanged nitrogen. He then investigated the reaction between pure oxygen and hydrogen, and confirmed the quantitative production of water. Cavendish's observations are first-rate, but he is cautious with the theoretical interpretation of his findings. He still took water to be the element, not hydrogen or oxygen, and believed that hydrogen gas was water united to phlogiston, and oxygen gas was water lacking phlogiston. When the two met, phlogiston was taken from the hydrogen (thereby forming water) and given to the oxygen (forming more water). He writes: 'We must allow that dephlogisticated air is in reality nothing but dephlogisticated water, or water deprived of its phlogiston; or, in other words, that water consists of dephlogisticated air united to phlogiston; and that inflammable air is either pure phlogiston, as Dr PRIESTLEY and Mr KIRWAN suppose, or else water united to phlogiston; since, according to this supposition, these two substances united together form pure water.'[57]

To whom the realization that water is not an element but actually composed of the elements hydrogen and oxygen should be attributed caused some debate lasting many decades. The issue was muddied by Cavendish's delay in publishing, since in 1783—a year before his interpretation appeared in print— James Watt, the brilliant Scottish mechanical engineer famed for his work on steam engines, had made known his interpretation of the experiment. Watt learned of the reaction from Priestley, and in April 1783, wrote to his friend, the Edinburgh chemist Joseph Black, to give an account virtually the same as Cavendish's.[58] Watt may have been more forceful in suggesting that water was a compound and not an element, but his interpretation was still in terms of phlogiston.

Cavendish notes in the published form of his paper, read in January 1784, that Lavoisier would interpret these results without phlogiston and state that water is composed of the two gases combined. However, he reports that when Lavoisier was informed of his experiments in the summer of 1781, the Frenchman had difficulty accepting the outcome. 'During the last summer also, a friend of mine gave some account of them to M. LAVOISIER, as well as of the conclusion drawn from them, that dephlogisticated air is only water deprived of phlogiston; but at that time so far was M. LAVOISIER from think-ing any such opinion warranted, that, till he was prevailed upon to repeat the experiment himself, he found some difficulty in believing that nearly the whole of two airs could be converted into water.'[59]

Lavoisier was incredulous about this experiment because he believed com-bination with oxygen must produce an acid, which is clearly not the case here. Indeed, he had tried on previous occasions to show that an acid was formed on burning hydrogen. Not only did he *not* succeed in producing any acid, he also missed the fact that water *was* produced. However, on 24 June 1783, Lavoisier tested Cavendish's experiment by burning hydrogen and oxygen from a blow-pipe in a closed glass vessel and collecting the water formed. In the report pub-lished the following year he writes: 'we do not hesitate to conclude that water is not a simple substance, and that it is composed, weight for weight, of inflam-mable air and vital air'.[60]

The fact that this inflammable gas enters into the composition of water sug-gested the name now in use for this element, proposed in a sweeping reform of chemical nomenclature that was to shake the chemical community.

A Method of Chymical Nomenclature

In the second half of the eighteenth century, many chemists were becoming increasingly dissatisfied with how the substances they used were named. A few individuals, notably the great Swedish mineralogist and chemist Torbern Bergman (1735–84), had made some preliminary proposals for improvement, but the big-gest reform was initiated in 1782, when Louis-Bernard Guyton de Morveau (1737–1816), a distinguished scientist from the provinces in France, published a paper translated as 'Memoir upon Chemical Denominations, the Necessity of Improving the System, and the Rules for Attaining a Perfect Language'.

In this paper Guyton presented a strong argument in favour of reform, and then proposed a series of rules or guidelines by which it could be achieved. These included: that substances should be referred to by meaningful names rather than descriptive phrases; that the names should not encourage erroneous ideas; that where possible the names should be derived from 'the dead languages'; and that the system should be adaptable to different modern languages.

Guyton anticipated his critics, and skilfully disarmed them by inviting suggestions and improvements. Not surprisingly, he soon came to the attention of Lavoisier and his circle. He travelled to Paris, and rapidly became a convert to Lavoisier's theories concerning oxygen and combustion. In 1787, Guyton, Lavoisier and two colleagues, Claude Louis Bertholet (1748–1822) and Antoine François de Fourcroy (1755–1809), published the book that was to mark the beginning of modern chemistry: the *Méthode de Nomenclature Chimique*. This was translated the following year into English as the *Method of Chymical Nomenclature*, and was based on Guyton's original paper, but with the ideas of Lavoisier incorporated. Their nomenclature was further reinforced with the publication two years later of Lavoisier's classic textbook *Traité Élémentaire de Chimie*, which first appeared in English as *Elements of Chemistry* in 1790.

Their system was to be based on a nomenclature which reflects the composition of substances, but in order to do that they needed to know, as far as possible, the ultimate components—the 'principles' or 'elements'. Perhaps, unlike their predecessors, they appreciated that what were then thought to be elements might not prove to be so in the future; for example, they strongly suspected this to be the case for the alkalis and silica. They write:

> We shall content ourselves here with regarding as simple all the substances which we cannot decompose; all such as we obtain in the last result from chymical analysis. Without doubt, in time to come, these substances which to us appear to be simples, will in their turns be decomposed, and probably we are at this epoch in respect to the siliceous earth and to the alkalies; but our imagination ought not to anticipate the facts, and we must not take upon us to say more than nature presents to our understanding.[61]

The 'Table of Simple Substances' from Lavoisier's *Traité* is shown in Figure 33. The first two entries in the table—light and heat—do not feature in any modern lists of elements, since we now know they are forms of energy. Lavoisier included them in his list because he thought they were components that made up matter. After all, heat and light were given out when things burned in air,

	Noms nouveaux.	Noms anciens correspondans.
Substances simples qui appartiennent aux trois règnes & qu'on peut regarder comme les élémens des corps.	Lumière.........	Lumière.
	Calorique........	Chaleur.
		Principe de la chaleur.
		Fluide igné.
		Feu.
		Matière du feu & de la chaleur.
	Oxygène........	Air déphlogistiqué.
		Air empiréal.
		Air vital.
		Base de l'air vital.
	Azote...........	Gaz phlogistiqué.
		Mofete.
		Base de la mofete.
	Hydrogène.......	Gaz inflammable.
		Base du gaz inflammable.
Substances simples non métalliques oxidables & acidifiables.	Soufre..........	Soufre.
	Phosphore.......	Phosphore.
	Carbone.........	Charbon pur.
	Radical muriatique.	Inconnu.
	Radical fluorique..	Inconnu.
	Radical boracique..	Inconnu.
Substances simples métalliques oxidables & acidifiables.	Antimoine.......	Antimoine.
	Argent..........	Argent.
	Arsehic.........	Arsenic.
	Bismuth.........	Bismuth.
	Cobolt..........	Cobolt.
	Cuivre..........	Cuivre.
	Etain...........	Etain.
	Fer.............	Fer.
	Manganèse.......	Manganèse.
	Mercure.........	Mercure.
	Molybdène.......	Molybdène.
	Nickel..........	Nickel.
	Or.............	Or.
	Platine.........	Platine.
	Plomb..........	Plomb.
	Tungstène.......	Tungstène.
	Zinc...........	Zinc.
Substances simples salifiables terreuses.	Chaux..........	Terre calcaire, chaux.
	Magnésie........	Magnésie, base du sel d'Epfom.
	Baryte..........	Barote, terre pesante.
	Alumine........	Argile, terre de l'alun, base de Palun.
	Silice..........	Terre filiceufe, terre vitrifiable.

Fig. 33. The 'Table of Simple Substances' or elements as drawn up by Lavoisier in his *Traité Élémentaire de Chimie*, published in 1789.

which suggested they must be contained either in the burning substance or in the air. The phlogistians thought the heat, in the form of phlogiston, was contained in all flammable substances; Lavoisier thought the heat—what he called 'caloric'—was in the air, specifically in the oxygen of the air. This is an important point since it influenced Lavoisier in his choice of name for oxygen.

Oxygen is the first element as we understand it included in Lavoisier's list. When discussing the name for this element in their book on nomenclature, the authors begin by discussing the names oxygen gas had received when first discovered: 'When the appellation of dephlogisticated air was changed into that of vital air, a choice more agreeable to reason was made, by substituting to an expression founded upon mere hypothesis, a term derived from one of the most remarkable properties of that substance, and which is so essentially characteristic of it that one should never hesitate to use it whenever it be necessary to indicate simply that portion of the atmospherical air which maintains respiration and combustion...'[62]

We might wonder why the authors did not therefore choose a name for the life-sustaining gas based on this essential, unique property. The reason was that Lavoisier thought oxygen gas, as present in the air, was not an element but was composed of his 'simple' caloric combined with another simple—the 'basis' of oxygen gas. They write: 'it is at present well demonstrated that this portion of the atmospherical fluid [oxygen gas] is not always in the state of air or gas, that it is decomposed in a great many operations, and loses, at least in part, the light and caloric which are what principally constitute it vital air'. In their eyes, this substance, the basis of vital air, demanded its own name. They chose the word 'oxygen', 'deriving it as Mr Lavoisier proposed, from the Greek words οξυς *acid*, and γείνομαι *I beget*, on account of the property of this principle, the basis of vital air, to change a great many of the substances with which it unites into the state of acid, or rather because it appears to be a principle necessary to acidity'.[63]

Lavoisier's idea that gases were the combination of caloric with the basis of other substances also influenced his naming of hydrogen. The authors write:

On applying the same principles to the aeriform substance called inflammable gas, the necessity of having a more explicit appellation is evident at the first view; it is true that this fluid is capable of being consumed, but this property does not exclusively belong to it, notwithstanding that it is the only substance which produces water by its combustion with oxygen gas. This is the property which appeared to us to be the most worthy of affording a name, not for the gas itself, which is a composition, but for the more fixed principle which constitutes the basis, and we have therefore called it *hydrogen*, from ὑδως *water* and γεινομαι *I beget*; experiments having proved that water is nothing but oxygenated hydrogen, or the immediate production of the combustion of oxygen gas with hydrogen gas, deprived of the light and caloric which disengage during the combustion.[64]

Sharp Chins and Vinegar Merchants

Lavoisier was soon made aware that his derivations of the words 'hydrogen' and 'oxygen' from the Greek for 'water-former' and 'acid-former' were somewhat questionable.[65] In the second issue of his book, which appeared in the same year as the first, he adds a footnote which appears in the first English translation of 1790:

> This expression Hydrogen has been very severely criticised by some, who pretend that it signifies engendered by water, and not that which engenders water. The experiments related in this chapter prove, that, when water is decomposed, hydrogen is produced, and that, when hydrogen is combined with oxygen, water is produced: So that we may say, with equal truth, that water is produced from hydrogen, or hydrogen is produced from water.[66]

In the second English edition, the translator adds that he is 'not Grecian enough to settle the grammatical dispute',[67] but the Irish physician Stephen Dickson holds no punches in further criticizing the derivations in his 'Essay on Chemical Nomenclature' in 1796. First, Dickson points out that the spelling proposed by the French, *oxigene*, suggests that the word stems from the Greek word for a 'cruet'—'οξις *(oxis)*'—and recommends that it would be better spelt as 'oxygen'.[68] However, he is still unhappy with this modification, saying, 'it is an old word usurped in a new and unwarrantable signification'.[69] He even writes, 'The Greek word οξυγενυς [oxygenys], which may be rendered in French or English oxygen, properly signifies *sharp chin*.' He writes that the Greek 'οξυς [oxys] literally signifies sharp or acute, and does not signify acid, except figuratively', and thinks the closest literal meaning of oxygen would be 'sharp-descended, or sprung from an edge'.[70] He does acknowledge that the 'oxys' part could suggest acid, but this does not help much: 'But οξυς, it may be said, was used by the Greeks, in combination, to signify acid also. It was so, though indeed very rarely, not in above five or six instances, for it generally indicated some other metaphorical meaning, or else vinegar. These few instances however, I grant, are sufficient to form a precedent: but what then? Oxygen, after all, signifies the *descendant* of an acid; whereas the inventors of this name intended and announced it to signify the *begetter* of acids.'

Along a similar line of reasoning, the staunch phlogistian Balthazar-Georges Sage wrote in 1800, 'If I don't use the *word oxygen*, it is because it signifies *son of vinegar merchant* [*fils de vinaigrier*].'[71]

After further detailed discussions, Dickson concludes that if the authors really wanted to denote 'begetter of acids', then 'The nomenclators should therefore, undoubtedly, have denominated this principle *oxygon*, not *oxigene*.'[72]

A more serious criticism that Dickson then raises is whether the name would be appropriate even if it did mean what Lavoisier had intended it to, since oxygen does *not* always form acids when it reacts with other species. 'Do the embraces of this begetter of acids always prove fruitful? We are told that oxigene is united to the matter of fire without generating an acid, to the basis of water without generating an acid, to most of the metals without generating an obvious and evolved acid; nay, that when superadded to acids themselves, it despoils them of part of their acidity.'

To be fair to him, Lavoisier did recognize that oxygen didn't always form acids when it reacted, but he did think it was the essential component common to all acids. Sadly, this was shown not to be the case when muriatic acid (later known as hydrochloric acid) was proved to consist only of chlorine and hydrogen. Later in the nineteenth century, the key component of acids (at least those in aqueous solution) was found to be hydrogen ions, and the pH scale was developed to measure the concentration of these ions in solution. Since hydrogen is therefore the basis of all acids, perhaps it would have been more appropriate to have named this element oxygen (or even 'oxygon'). The real unique property of real oxygen is that it combines with hydrogen to form water, so perhaps the name hydrogen (or 'hydrogon'?) would have been appropriate for this element. In fact, even in 1788, it was suggested that the name hydrogen would be better suited to the element oxygen, since the latter makes up the greatest part by weight of water. In an article discussing the new nomenclature, the Spanish author Don Juan Manuel de Arejula writes: 'The name of hydrogen...is as improper as its antecedent; for if the name of hydrogen means "engenderer of water", this is better suited to oxygen, since to form a quantity of water you need five and a half parts of the latter, and one of the former by weight; or as proposed by Lavoisier, fifteen grains of flammable gas, and eighty-five of vital air make up a hundred of water.'[73] We now know that the correct combining ratio would be one part (by weight) of hydrogen to eight parts of oxygen, so while the detail was not quite correct, the principle was sound.

It would therefore be more logical if the names for hydrogen and oxygen were swapped round, and perhaps we should be writing the formula for water as O_2H rather than H_2O. Nonetheless, Lavoisier's names stuck and are now firmly established; not so his proposal for the element we know as nitrogen, which was swiftly abandoned in most languages.

Alkaligen, Azote, and Septon

Finding a name for the main component of air—nitrogen—proved more problematic for the French authors. In their *Method of Chymical Nomenclature*, they outline why the phrase 'phlogisticated air' is far from ideal since it suggests that normal air can be modified to produce the gas, whereas they believe (correctly) that it is an element in its own right.

Since it was believed to be in certain alkalis they considered calling it 'alkaligen', but dismissed the idea because it had not been established whether it was the essential component of *all* alkalis. In Lavoisier's *Elements* he adds, 'beside, it is proved to compose a part of the nitric acid, which gives as good reason to have called it *nitrigen*'.[74] The four authors concluded that 'it was not possible by a single word to express the double property of forming the radical of a certain acid, and assisting in the production of an alkali', and so they went for a safer option:

> In this situation we thought it were the better way to derive the denomination from its other property, which it manifests in a very great degree, viz. not to maintain the existence of animals, to be really non-vital; in short to be so in a more considerable respect than the hepatic and acid gases, which do not like it constitute an essential part of the atmospherical mass, and therefore we have denominated it *azot* from the Greek privative α and ζωή *life*. After this it may not appear difficult to remember that the air which we breathe is a composition of oxygen gas and azotic gas.[75]

This term, '*azote*', is still in common use in France for the element nitrogen. Although it did not survive long in English, its ghost persists, with the terms 'azo' and 'azide' being used for certain arrangements of nitrogen atoms in molecules. Dickson discusses the alternative names given to nitrogen and writes of the movement away from the term 'mephitic air' with his typical dramatic flair:

This term [*mephitic*], it would seem, did not appear sufficiently novel to the confederate reformists; for although it had borne sway on the continent for some time, it was held meet, at the grand revolutionary council of nomenclature in Paris, that it should be formally deposed, and *azotic gas* erected in its stead. Philology revolts against this monstrous combination of heterogeneous sounds, dragged from ancient Greece, and the modern United Provinces, and tied together in despite of taste and judgment.[76]

He adds: 'Azotic gas, strictly interpreted, signifies *a lifeless emanation of chaos*.' However, the general dissatisfaction of the wider chemical populace was more reasonable—it was pointed out that *all* gases with the exception of oxygen were fatal to life, and so it was hardly appropriate to name one gas because of this lethal property. This was first spelt out by one of the first adopters of the new chemistry, the French chemist Jean Antoine Chaptal (1756–1832), who in 1790 wrote that azotic gas was inappropriate 'because, none of the known gaseous substances excepting vital air being proper for respiration, the word Azote agrees with every one of them except one; and consequently this denomination is not founded upon an exclusive property, distinctive and characteristic of the gas itself'.[77]

He also pointed out that there were compounds, such as nitre and nitrous acid, that contained the element, but their well-established names did not currently reflect this fact; if the name azote were to be kept, nitrous acid should become 'azotic acid'. Rather than change the names of the compounds, he simply proposed to change the name of the element azote: 'I have presumed to propose that of Nitrogene Gas. In the first place, it is deduced from the characteristic and exclusive property of this gas, which forms the radical of the nitric acid. By this means we shall preserve to the combinations of this substance the received denominations, such as those of the Nitric Acid, Nitrates, Nitrites, &c.'[78]

Nitrogen was not the only alternative suggested for azote. Samuel Latham Mitchill, professor of chemistry, natural history, and agriculture at the College of New York, proposed different names for compounds of nitrogen, and for the element itself. He writes:

It is a pity, that notwithstanding all these things, the French Academicians who framed the new Nomenclature, suffered themselves to retain the words *nitrous* acid, *nitrous* gas, &c. which seem to me to be very improper, and to be quite as subject to objection as the terms *azote* and *nitrogene*, for their radical. The mind becomes unhappily impressed with the notion of those products being derived from nitre, whereas the fact is, nitre derives its origin from this *animal acid*. Had I been a member of that committee of that academy, I should have proposed to derive the name of the radical from the Greek verb σηπω [sepo], *putrefacio*; to call it σηπτον, *putridum*; and have the made the Nomenclature stand thus:

1 Septon; instead of azote or nitrogene.
2 Septous gas; instead of azotic gas or nitrogene gas.
3 Gaseous oxyd of speton; instead of gaseous oxyd of azote or of nitrogene.
4 Septic gas; instead of nitrous gas.
5 Septous acid; instead of nitrous acid.
6 Septic acid; instead of nitric acid.
7 Septate; Septite, &c &c.[79]

In some ways it's a shame we didn't end up with the word 'septon' for nitrogen— since on first hearing, it sounds like it derives from the Greek for seven, and we now know that nitrogen is unique because of the seven protons it has in its nucleus. 'Septic acid', on the other hand, doesn't have such a pleasant ring to it. Ideas concerning the putrefaction of animal matter in the formation of nitrates will be explored further in the following chapter, but for now, we shall simply note that it was the term 'nitrogen', so close to the *'nitrigen'* rejected by the French authors, that became generally accepted and is now the official English name for the element.

Kept in Translation

Much of the language initially suggested by the four French authors remains with us today. In addition to the names of the elements hydrogen and oxygen, they also proposed a system for substances composed of two elements, such as the oxides, and for so-called radicals consisting of groups of atoms that stay together during many reactions: sulfate (consisting of sulfur and oxygen), carbonate (carbon and oxygen), nitrate (nitrogen and oxygen), and phosphate (phosphorus and oxygen) are examples that we still use today.

Some countries, such as England and Spain, essentially took the new words, invented by the French but based on classical origins, and adapted them only slightly to the quirks of their own languages. Others, such as Germany, took the ideas behind the names and translated those into their own languages. The words for hydrogen and oxygen in German, *'wasserstoff'* and *'sauerstoff'*, literally mean 'water-stuff' or 'water-matter' and 'acid-stuff' or 'acid-matter'. Lavoisier's ideas have even ended up in the Japanese names for these elements. The ideas from Western 'modern' chemistry were not introduced there until the nineteenth century, when suddenly many new words and concepts needed to be created all at once. The first Japanese textbook of modern chemistry is often taken to be the *Seimi kaiso* from 1837, by Udagawa Yoan (1798–1846). This was based on a number of Dutch books, which were themselves translations of other works. When needing to create names for the new elements and concepts, Udagawa preserved the meaning behind the names used in the Dutch works, which ultimately came from Lavoisier: his words for hydrogen and oxygen, which are still used today, are '*suiso*' ('water element') and '*sanso*' ('acid element'). This means the term 'acid element' was introduced for the element oxygen at a time when it was realized that it was not, after all, the key principle of acids.

In the early nineteenth century, Hungarian chemists attempted to introduce into their language a new patriotic system of element names based on their unique properties. The suggested word for oxygen was '*éleny*', which may be translated as 'the element which keeps you alive' and is reminiscent of the earlier 'vital air'. Sadly, the system did not catch on, and they now use the word '*oxigén*'. Only the Chinese seem to have sensibly used a name for oxygen with its origins based on the life-sustaining properties of the gas. The character they use consists of components which may be interpreted as 'nourishing gas', although confusion is caused by part of the character being used to signify 'sheep'—a nourishing foodstuff—which is why one occasionally hears that the character for oxygen means 'sheep-gas'.

Trace of Scheele's term for oxygen, 'fire-air', seems to linger only in Danish and Lithuanian. As in Hungary, these countries also wanted names based on their own language. In 1814, Hans Christian Ørsted introduced a proposal for new Danish terms in chemistry after noting that the language then in use was merely a translation of the French antiphlogistic terms. He suggested the words '*Ilt*' and '*Brint*' to replace '*Suurstof*' and '*Vandstof*', then in use for oxygen and hydrogen respectively. '*Ilt*' he derived from the word for 'fire', and '*Brint*' from '*at braende*', meaning 'to burn'. While these terms are still in use in everyday Danish, 'oxygen' and 'hydrogen' are used in scientific contexts. Similarly, the Lithuanian linguist Jonas Jablonskis introduced many new words into the language in the early twentieth century in a drive to reinforce a national identity. His word for oxygen was '*deguonis*', derived from '*degti*', meaning 'to burn'.

Phlogiston and Caloric R.I.P.

We have seen how chemists soon criticized the new language proposed by Lavoisier, but some were also reluctant to accept his theoretical views. In 1788, just one year before Lavoisier brought out the famous textbook containing his ideas and nomenclature woven together in a course of chemistry, the English physician John Berkenhout wrote: 'I have mentioned a new sect of philosophers called *Antiphlogistians*. The reader may perhaps be curious to know something of their origin and their creed. Who was the real founder of this sect, I am not quite certain. I think that honour is due to M. Lavoisier, a Chemist high in fame, and very deservedly so: but, I believe, their whole system was first promulgated in Fourecroy's [sic] *Elements de Chimie*.'[80]

After giving an excellent summary of Lavoisier's ideas, which to this day are some of the basic principles of modern chemistry as taught to young students beginning the subject, Berkenhout concludes: 'Such are the fundamental principles of this new philosophical Chemistry. It was born in France, and there let it die. It has been considered in other nations only to be ridiculed.'[81]

Resistance to the new ideas in Germany was perhaps initially fuelled by the perceived relegation of their master chemist Stahl. An interesting and quite bizarre account of this from 1794, by a respected professor from Gottingen, is included in the edition of Edinburgh chemist Joseph Black's *Lectures on the Elements of Chemistry*, published posthumously in 1803:

> Great hesitation, doubt, and objections, were to be expected in Germany, the native soil of chemistry, and the resort of all who wished to perfect themselves in mineralogy. The new doctrines were even received with aversion and disgust. This, he says, was chiefly owing to the character of the nation from whence they came. The Germans, who had been accustomed to consider themselves as the chemical teachers of Europe, could not bear to hear the opinions of their master, Stahl, treated with contempt; to be told by Frenchmen, living among them for instruction, that the principles of Stahl were such as no man could embrace who had a spark of common sense; to be told, in letters from France, that the principle of Stahl was a *mera qualitas*; a *mera contemplatio*, a fancy of the brain, which disgraced any man who entertained it for a minute; and to have it added, with saucy politeness, *dulci requiescat in pace!* But what most provoked them, was the pitiful triumphs of victory in which the French chemists indulged themselves. He says, that when the association had finished their experiments on the composition and decomposition of water, which filled up all the gaps of the system, they had a solemn meeting in Paris, in which Madame Lavoisier,

in the habit of a priestess, burned on an altar Stahl's *Chemiae dogmaticae et Experimentalis Fundamenta*, solemn music playing a *requiem*; and he remarks, that if Newton had been capable of such a childish triumph over the vortices of Des Cartes, he could never be supposed the man who wrote the *Principia*. I might add, that if Newton or Black had so exulted over Des Cartes and Meyer, their countrymen would have concluded that they were out of their senses. But at Paris every thing becomes a mode, and must be *fêté*.[82]

It is interesting that the idea of phlogiston was not made entirely obsolete by Lavoisier's revelations—rather than the 'combustible substance' contained 'phlogiston' in the old theory, while the 'oxygen gas' contained 'caloric' in the new. Remember, light and heat were the first two entries listed in Lavoisier's 'Table of Simples'. In the preface to the English translation of the *Nomenclature*, the translator comments:

> For though the late experiments demonstrate that phlogiston does not give weight or heaviness to metals, that phlogiston does not disengage itself from the sulphur during the formation of the sulphuric acid; yet we still allow the absolute existence of a phlogiston. It is still the matter of fire, of flame, of light, and of heat, which is liberated in combustion; the only difference is, that we do not agree with Stahl, that this principle disengages from the body in combustion; but by a variety of concurring experiments are induced to believe, are convinced, that it is liberated from the vital air on the precipitation of the oxygen. Yet it is still phlogiston with its most distinguishing attributes. In short, it is still the matter of heat; whether we call it phlogiston, caloric, or in plain English, *fire*. But whether light, and heat, or that which gives the sensation of heat, be essentially different bodies, or only modifications of the same, is not at all determined.[83]

Eventually, Lavoisier's caloric theory was completely disbanded when light and heat were recognized as forms of energy, and oxygen gas was shown not to be a compound of oxygen and caloric. But before this, an interesting modification of the theory was proposed by a bold twenty-year-old Cornishman who even suggested a new name for oxygen. He states:

> Oxygen gas, (which the French nomenclators have assumed to be oxygen combined with caloric) will be proved to be a substance compounded of light and oxygen. It would be highly improper to denote this substance by either of the terms oxygen gas, or oxygen. The one would signify that it was a simple substance combined with caloric, the other that it was a simple substance, the

acidifying principle. The term *phosoxygen* (from φως light, οξυς acid, and γενητορ generator) will I think be unexceptional; it will express a chemical combination of the simple substance light, with the simple substance oxygen; it will not materially alter the nomenclature of the French philosophers; and as will be seen hereafter, it can be easily modified to express, in conjunction with other words, the combinations of light and oxygen.[84]

In addition to 'phosoxygen', the author suggests 'phosoxyds' to replace Lavoisier's 'oxyds', 'phosnitrates' to replace 'nitrates', and so on. Of course, these suggestions never caught on. Within a year, with tail between his legs, the rash youngster wrote: 'I beg to be considered as a sceptic with regard to my own particular theory of the combination of light and theories of light in general. On account of this scepticism, and for other reasons, I shall in future use the common nomenclature.'[85]

Despite this initial setback, the lad went on to become one of Britain's most famous chemists, Sir Humphry Davy. While missing the mark with his phosoxygen, it was he who later proved that muriatic acid contained no oxygen (just hydrogen and the element he named chlorine), providing the final death blow to Lavoisier's oxygen theory of acidity. As we shall see in the next chapter, it was also Davy who was the first to show what Lavoisier had suspected: that the alkalis soda and potash were not elements, but could be further 'decomposed'. And, through the invention of the famous Davy Lamp, he was responsible for saving many lives by providing safe illumination in the mines without risking the explosion of the dreaded 'fire-damp'.

5

OF ASHES AND ALKALIS

With water I bathed myself. With soda I cleansed myself.
—From an ancient Sumerian clay tablet,
third millennium BC

The name azote, proposed by Lavoisier and his colleagues, did not gain wide acceptance; nitrogen, meaning 'nitre-former', is the name now familiar to us. Modern chemists understand 'nitre' to mean 'potassium nitrate', one of the key ingredients of gunpowder, containing the elements potassium, oxygen, and nitrogen. However, although it dates back to antiquity, the name nitre initially referred to a completely different compound containing no nitrogen at all. It is the Latinized name, *natrium*, derived from this original use, that gives us the modern chemical symbol Na, for the element Humphry Davy named sodium.

Ancient Nitre

Travellers to modern-day northern Egypt may find themselves in a region known as the Nitrian Desert, or the Natron Valley—*Wadi El Natrun*. Here, ancient Egyptians would collect crude salt mixtures from certain lakes and use them for a variety of purposes, such as cleaning, making glass, embalming, and the preparation of medicines. The Egyptian word for the salt may be written 'ntry' or 'ntr' ('neter'), and it has survived for over three thousand years through variations including 'neter' (Hebrew), 'nitron' (Greek), 'nitrum'(Latin), and more modern modifications 'nether', 'niter', 'nitre', 'natrun', and 'natron'. Bartholomeus Anglicus, the thirteenth-century monk and author of *De proprietatibus rerum* ('On the Properties of Things'), quotes Isidore of Seville from five hundred years earlier saying: '*Nitrum* hath ye name of the countrey of *Nitria* that is in *Aegypt*. Thereof is medicine made, & there with bodies and clothes be cleansed and washed.'

Whether the salt was actually named after the region or vice versa is not clear. Although its composition varied enormously, what distinguished nitre from common salt was the presence of significant proportions of sodium carbonate and sodium bicarbonate (sodium hydrogen carbonate). In addition to these carbonates, analyses of ancient samples, including that used in the embalming of the pharaoh Tutankhamun, who died in 1352 BC, also reveal large proportions of common salt (sodium chloride), sodium sulfate, and silica (silicon dioxide), with smaller proportions of calcium and magnesium carbonates and other minor impurities.

That early nitre really was a form of impure sodium carbonate, and distinct from our modern potassium nitrate, is supported by the occurrence of the Hebrew *neter* in the Bible (translated as 'nitre' in the King James version of 1611), where there is a reference to a key reaction characteristic of carbonates but not of nitrates. With dilute acid, carbonates effervesce, rapidly giving bubbles of carbon dioxide—it has even been suggested that the Hebrew *neter* derives from *natar*, meaning to effervesce. The reaction is referred to in Proverbs 25:20: 'As hee that taketh away a garment in cold weather; and as vineger upon nitre; so is he that singeth songs to an heavy heart.' A modern interpretation of this verse suggests that singing songs to a sad person is as foolish as taking off a coat on a cold day, or mixing soda and vinegar (acetic acid), which would destroy the effect of the soda.

Nitre appears again in Jeremiah 2:22, where reference is made to its use in washing: 'For though thou wash thee with nitre, and take thee much sope, yet thine iniquitie is marked before me, saith the Lord GOD.'

This use of nitre for washing also features in one of the proverbs collected by the fifteenth-century scholar Erasmus of Rotterdam. '*Asini caput ne laves nitro*' may be translated as 'don't wash the donkey's head with soda'—implying it is a waste of good detergent. Pure sodium carbonate may still be purchased for cleaning purposes, but it is now sold under the name washing soda rather than nitre.

Pliny gives an account of how the Egyptians prepare *nitrum* by the evaporation of water from the Nile, in much the same way that common salt may be prepared by evaporating sea water. The following English translation is by Philemon Holland, Doctor of Physicke, who in 1601 translated Pliny's 'nitrum' as 'nitre'.

> As for artificiall Nitre, great abundance there is made of it in Ægypt, but farre inferiour in goodnesse to the other: for browne and duskish it is, and besides full of grit and stones. The order of making it, is all one in manner with that of

salt, saving only that in the salt houses they let in sea water, wheras into the boiling houses of Nitre they conveigh the water of the river Nilus.[1]

Pliny's description was picked up in the 1556 work *De re metallica* by Agricola, who also included an imaginative woodcut to illustrate how he understood the process (Figure 34). Agricola writes:

Nitrum is usually made from *nitrous* waters, or from solutions or from lye. In the same manner as sea-water or salt-water is poured into salt-pits and evaporated by the heat of the sun and changed into salt, so the *nitrous* Nile is led into *nitrum* pits and evaporated by the heat of the sun and converted into *nitrum*. Just as the

Fig. 34. A woodcut from the 1556 work *De re metallica* by Agricola. Water from the River Nile flows into the nitrum pits (B), from where the nitre crystals are collected after evaporation of the water.

sea, in flowing of its own will over the soil of this same Egypt, is changed into salt, so also the Nile, when it overflows in the dog days, is converted into *nitrum* when it flows into the *nitrum* pits.[2]

Both Pliny and Agricola are mistaken in their accounts in that it is not the Nile water itself that contains the precious salts, but the earth in specific regions. During the flooding of the Nile, the salts are leached out, and the resulting solutions form small lakes from which the salts may be collected as the water evaporates.

Glass

In addition to its use as an early detergent, the alkaline nitre was also used extensively in the manufacture of glass. By 2500 BC glass was being produced in various regions in Mesopotamia and Egypt, though isolated samples have been found from far earlier. These early productions were not like our modern glass; transparency was not valued, and samples were largely used as imitations of precious stones. The key ingredients in the most common glass (now called soda-glass or, more correctly, soda-lime glass) are silica (or silicon dioxide from sand), soda (sodium carbonate), and lime (from calcium carbonate). In the very high temperatures of the glass furnace, both the sodium carbonate and calcium carbonate decompose to the oxides, giving out their 'fixed air' (carbon dioxide). The role of the sodium carbonate is to act as a flux—something that helps the silica melt at lower temperatures than it would otherwise do. Derived from the Latin *'fluxus'* meaning 'flow', fluxes were probably first discovered during the process of metal smelting. But heating silica and sodium carbonate without the calcium carbonate does not give a stable glass; at best it might give a crude glass that decays in the presence of moisture. At worst it can form sodium silicate, a substance also known as water glass, which has the remarkable property of being soluble in water to yield a thick, syrupy liquid that was once used to preserve eggs. The formation of this 'soluble glass' was described in the first half of the seventeenth century by van Helmont (he of the 'gas' fame): 'Moreover, Stones, Gemmes, Sands, Marbles, Flints, &c. through an *Alcali* being joyned unto them, are glassified: but if they are boyled with the more *Alcali*, they are indeed resolved into moisture.'[3]

Most salts of sodium are soluble in water and in order to create a stable glass, less soluble components are needed. This is the purpose of adding lime, which is said to act as a stabilizer. Since the purity of both the sand and the soda used in early glassmaking processes was poor, calcium carbonate, for example, in the form of pulverized shells, was inevitably present in the mixture, and so the necessity of the lime was not recognized until the late seventeenth century.

Pliny relates a story concerning the supposed discovery of glass from the particularly fine sand found where the river Belus flows into the sea, close to the settlement of Ptolemais in Syria. He says:

> The coast along this river which sheweth this kind of sand, is not above halfe a mile in all, and yet for many a hundred yeare it hath furnished all places with matter sufficient to make glasse. As touching which devise, the common voice and fame runneth, That there arrived sometimes certain merchants in a ship laden with nitre, in the mouth of this river, and being landed, minded to seeth their victuals upon the shore and the very sands: but for that they wanted other stones, to serve as trevets to beare up their pans and cauldrons over the fire, they made shift with certaine peeces of sal-nitre out of the ship, to support the said pans, and so made fire underneath: which being once afire among the sand and gravell of the shore, they might perceive a certaine cleare liquor run from under the fire in very streams, and hereupon they say came the first invention of making glasse.[4]

The Vegetable Alkali

Analysis of glasses from the fourteenth century BC reveals that Egyptian nitre was not the only source of sodium carbonate used; another was obtained from burning plant matter. Pliny noted that nitre used to be prepared in this way by burning oak wood: 'In times past men have practised to make Nitre, of Oke wood burnt; but never was there any great store thereof made by that devise: and long it is since that feat was altogether given over.'[5]

Burning wood later became a major source of alkaline carbonates, but before this, the ashes of certain herbs were found to be full of the valuable salts necessary to the glass maker. Such use of plant ashes seems to have originated in the Levant, an area including modern-day Israel and Palestinian territories, Jordan, Lebanon, and Syria. The Arabic name for the ash was *qaliy* or *kali*. With the definite article 'al' or 'the', *al-kali* gives us the word we still use today—'alkali'—the soluble substances which share the property of being able to neutralize acids. A description of the type of plants used for making these carbonate salts is found in the herbals of the sixteenth century:

> The herbe named of the Arabians Kali, or Alkali hath many grosse stalkes, of halfe a foote or nine inches long: out of them groweth small leaves, somewhat

long & thicke, not much unlike the leaves of Prickmadam, saving they be longer, and sharpe poynted, with a harde prickley toppe or poynt, so that for this consideration the whole plant is very rough and sharpe, and his leaves be so dangerous and hurtfull by reason of their sharp prickles, that they cannot be very easily touched...This herbe is salte and full of iuyce or sap...[6]

The author mentions other similar plants:

There is an other herbe in nature much like unto this, the whiche is called *Salicornia*...this plante is also salte in taste and full of iuyce like kali. Of these two plantes are made *Alumen Catinum*, and *Sal Alcali*, whiche is much used in the making of glasses, and for divers other purposes.

'*Alumen Catinum*' refers to the ashes of these plants, and is named after the Latin for the dish or bowl used in the drying process, *catinum*. *Sal Alcali* (alkaline salt) is a purified form of the alkali extracted from the ashes. Names given to these plants in other countries are also mentioned:

The first is called in Italian *Soda*: in Spanish *Barilla*, and *Soda Barilla*: and it is the right Kali, or Alkali of the Arabians: some call it in English Salte worte, we may also call it Kali, or Prickled Kali.

The second is now called *Salicornia*, & it is a certaine kinde of Kali. Some call it in English Sea grape, and knotted or ioynted Kali.

The Axsen or asshes, whiche are made of burnt Kali, is called in Latine of the Alcumistes and Glassemakers *Alumen Catinum*, but the Salte whiche is made of the same Axsen, is called *Sal Alcali*.

To make the *Sal Alcali*, first the ashes are mixed with water and then allowed to settle. The alkaline solution above the residual solids is known as lye, and this yields the solid alkali on evaporation. The process of treating the ashes with water was known as 'lixiviation', '*lixivium*' being the Latin for lye, coming from '*lix*', meaning 'ashes'. The process is described and illustrated by Agricola (Figure 35):

Lye is made from the ashes of reeds and rushes...The ashes, as well as the earth, should first be put into a large vat; then fresh water should be poured over the ashes or earth, and it should be stirred for about twelve hours with a stick, so that it may dissolve the salt. Then the plug is pulled out of the large vat; the solution of salt or the lye is drained into a small tub and emptied with ladles into small vats; finally, such a solution is transferred into iron or lead caldrons and boiled, until the water having evaporated, the juices are condensed into salt.[7]

Fig. 35. A woodcut from the 1556 work *De re metallica* by Agricola, illustrating the production of alkali. Salt-rich plants were burned and their ashes collected and dissolved in water to yield solutions of the alkalis.

The type of ash produced was often named after the plant itself—hence soda ash (also called *soude*) or the barilla ash particularly common in Spain. Even seaweed, or sea-wrack, was used, and the ashes were commonly known as 'kelp'. As well as the shrubs, wood later became a valuable source of alkaline ash, but with a different composition. Just as the composition of Egyptian nitre varied enormously, so too did the alkaline salts obtained from the ashes of plants. If the plant grew near the sea, as did many of the kali herb plants, the ash was more likely to contain sodium carbonate (the main component of the Egyptian nitre). When the plant grew away from salt water, as was the case with most trees, the ash contained more potassium carbonate.

The production of great quantities of alkali prepared from wood-ash in large pots gave rise to a new term as described in the early eighteenth century by Hermann Boerhaave:

> *Potas* or Pot-ashes…is brought yearly by the Merchant's Ships in great abundance from *Coerland* [now part of Latvia and Lithuania], *Russia*, and *Poland*. It is prepared there from the Wood of green Fir, Pine, Oak, and the like, of which they make large piles in proper Trenches, and burn them till they are reduced to Ashes…These ashes are then dissolved in boiling Water, and when the Liquor at top, which contains the Salt, is depurated [purified, freed from impurities] by standing quiet, it is poured off clear…This, then, is immediately put into large copper Pots, and is there boiled for the space of three days, by which means they procure the Salt they call *Potas*, (which signifies Pot-Ashes) on account of its being thus made in Pots.[8]

Our modern word 'potassium' derives from these 'pot-ashes', but its symbol, K, is from the Latinized word '*Kalium*', from the Arabic Kali.

A purer form of potash, albeit under a different name, could also be prepared from another product of plant origin: wine-lees, the white precipitate or dregs left in the bottle or at the bottom of the barrel (Figure 36). Known as 'tartar', the pure solid would now be called potassium hydrogen tartrate, and it is still sold as cream of tartar. This substance decomposes on heating, leaving behind, after purification, what was known as 'salt of tartar': potassium carbonate. In the *Royal Pharmacopoea, Galenical and Chymical* of 1678 by Charas, he states 'Salt of Tartar is as it were a model of all the fix'd Salts of Plants.'[9] He goes on to describe the vigorous reaction of this carbonate with acid (evolving much carbon dioxide) which he says many people take 'for an effect of the Antimony [discord] which they believe there is, between the Acids and the Alkales'.

Fig. 36. A woodcut from the *Ortus sanitates*, published in the late fifteenth century. The labourer is shown scraping the barrel for wine-lees.

Throughout Roman times, the main source of alkaline carbonates—certainly that used in the production of glass—was Egyptian nitre, but eventually that derived from plant sources came to dominate; kali, soda, or pot-ash were all used, with little distinction between them. For most early uses it made no difference whether the ash contained more sodium or potassium carbonate since the two substances are remarkably similar to each other and no chemical distinction between them was made until the eighteenth century. However, as Egyptian nitre became less commonly used, the term 'nitre' gradually became associated with a completely different salt. It probably came about because, like the ancient nitre, this salt could also be extracted from the earth, or found 'growing' as an efflorescence on certain undisturbed stones or bricks. In the sixteenth century, some authors used the term 'sal-nitre' (or similar variations) for this new salt in order to distinguish it from the ancient nitre of the Egyptians. This sal-nitre acquired a new importance with the proliferation of weapons requiring gunpowder.

Sal-nitre—Saltpetre

And that it was great pitty, so it was,
This villainous saltpeeter should be digd
Out of the bowels of the harmeles earth[10]

The discovery of gunpowder was not possible until its key ingredient—saltpetre (potassium nitrate)—was identified and could be routinely prepared. This first took place in China, where this substance came to be known as '*Hsiao*', which Joseph Needham in his monumental *Science and Civilisation in*

China translates as 'solve'. This meaning came about possibly from the use of potassium nitrate as a flux in metallurgy, or possibly from its use in the preparation of nitric acid, which could dissolve many metals and minerals. Needham quotes from a pharmacopoeia of the Khai-Pao Reign-Period, 973 AD:

> It was because saltpetre can dissolve and liquefy minerals that it was given the name of solve-stone... Solve-stone is in fact a 'ground frost', an efflorescence of the soil. It occurs among mountains and marshes, and in winter months it looks like frost on the ground. People sweep it up, collect it, and dissolve it in water, after which they boil to evaporate it, and so it is prepared... Actually solve-stone is produced among the rocks and cliffs in the mountains west of Mou-chou in Szechuan.[11]

The name occurs much earlier, dating back to the fourth century BC, but none of its most characteristic properties are mentioned, so it might not be referring to the same substance. However, in AD 492, Thao Hung-Ching clearly discusses the substance in his work translated as *Collected Commentaries on the Classical Pharmacopoeia of the Heavenly Husbandman*. In this is described the unmistakable purple flame test given by the solve-stone (characteristic of potassium compounds), together with the powerful flames it produces on hot charcoal.[12]

Although saltpetre had been known from an early age, its use in gunpowder was not discovered until much later, probably around AD 1040 (a couple of hundred years before its European 'discovery'). Before gunpowder, saltpetre was used in various recipes to make fire, and even medicines. This is reflected in the Chinese name for gunpowder, *huo yao*, which may be translated as 'fire-chemical', 'fire-drug', or 'fire-medicine'. In his *Great Pharmacopoeia* of 1569, Li Shih-Chen writes:

> Gunpowder has a bitter-sour sapidity, and is slightly toxic. It can be used to treat sores and ringworm, it kills insects, and it dispels damp chii and hot epidemic fevers. It is composed of saltpetre, sulphur and pine charcoal, and it is used for various preparations for beacon-fires, guns and cannon.[13]

The fact that the Western world only uses the word gunpowder supports the idea that this mixture was essentially imported ready-made for use in weapons, as opposed to having gradually evolved through different uses.

One of the earliest detailed accounts in Western literature of the preparation of the potassium nitrate used in making gunpowder occurs in an Italian work, *De la Pirotechnia*. The author, Vannoccio Biringuccio (1480–ca. 1539), was a master metallurgist and his work, published in 1540 shortly after his death, has been described as the earliest printed work to cover the whole field of metallurgy. In addition to the smelting of metals from their ores, he describes in great detail how the metal may be cast into cannon, guns, and other artillery, as well as less belligerent objects such as statues and bells. He includes detailed descriptions of the preparation of gunpowder, since he says if he failed to do so, 'it would be as if I had pointed out only the useless shadow of a thing'. He adds 'For these reasons I now wish to tell you, in addition to what I have already said, how saltpeter is made and the methods of refining it well; what it is and how without it guns and many effects of violent and artificial fires would have been discovered in vain.'[14]

When he first talks of different salts, Biringuccio mentions a natural nitre (he uses the word 'nitro') mined in the form of a mineral stone in Armenia, Africa, and Egypt, and prepared from nitrous waters. This nitre is the ancient carbonate mixtures discussed earlier. He then goes on to speak of artificial nitre: 'Artificial nitro has the same qualities but it is much more combustible than that which is called natural. The ingenious moderns have recognized that this exists in certain kinds of soil and with art they have found a way to extract it from them. They have called it not nitro but salnitro.'[15]

Although a full English translation of Biringuccio's work did not appear until 1942, the section on the preparation of saltpetre was 'borrowed' and appended to Peter Whitehorne's 1562 translation of Machiavelli's The Arte of Warre. Whitehorne, like Smith four hundred years later, translates Biringuccio's 'salnitro' as 'saltpeter', and writes:

> Saltpeter is a mixture of manie substaunces, gotten oute with fire and water of drie and durtie grownde, or of that flower, that growth owte of newe walles, in selars, or of that grownde whiche is fownde loose within toombes, or desolate caves, where raine cannot come in … but the moste excellenteste of all other, is made of the dunge of beastes, converted into yerthe, in stabells or in dunghills, of long time not used: and above all other, of the same that cumeth of hogges, the moste and best is gotten.[16]

The fact that it may be found 'growing' on walls fits in with the popular etymology of 'saltpetre': as Ercker stated in the sixteenth century, 'Salt-Petre is a Stone-Salt'[17] (Der Salpeter ist ein Steinsaltz),[18] saltpetre therefore suggesting a salt from rock (petros and petra being the ancient Greek and classical Latin for rock or stone). However, in English, rock salt itself is actually common salt—sodium chloride—that has been extracted from a mine (as opposed to having been recovered from sea water). It has been suggested that the word 'saltpetre' was essentially a derivative following '*sal-nitro*' in Italian, '*salitra*' in Romanian, '*salitre*' in Spanish and Portuguese, and the '*salniter*', '*salliter*', '*saliter*', '*salbeter*', and '*salpeter*' that appear in Middle High German.[19]

As it was the essential ingredient for gunpowder, it was crucial for any government to secure a steady supply of saltpetre. Before the discovery of vast nitrate reserves in India (and much later, Chile), saltpetre had to be produced domestically or, more commonly, in continental Europe. The much-loathed 'saltpetremen' were licensed to enter property in order to search for the precious salt. They were allowed to dig up any earth from cellars and stables suspected of containing traces of nitrates (Figure 37).

Attempts were also made to produce saltpetre artificially. Although the practice was not so common in England, continental saltpetremen constructed

Fig. 37. A saltpetreman at work, pulling up the floor in a stable to get to the earth beneath. An engraving from a broadsheet issued by a Swiss firework manufacturer in the early eighteenth century.

Fig. 38. Artificial nitre pits shown in a woodcut from Ercker's mining work from 1580.

nitre pits in which long, shallow mounds of earth were fed with excrement and lime (Figure 38). These helped the bacteria present to convert ammonia from the urine into calcium nitrate, from which the saltpetre could be prepared.

This production of nitre from the putrification of animal waste indicates why Prof Mitchill (1792–1801) of New York proposed for nitrogen gas ('nitre-former') the word 'septon', from the Greek for 'putrification', and the 'septic acid' derived from it.

The Rise of Natron

As the term 'nitre' gradually became more associated with saltpetre (potassium nitrate), a new variation of the word, 'natron', emerged for the original 'nitre' (sodium carbonate). After publication of Lemery's chemistry text in French in 1675, there must have been significant confusion as to the meaning of 'nitre', since it was necessary for the author to clarify in the second edition just two years later:

IT is probable that the *Niter* of the antients was either the *Aegyptian Natron*, or a salt that is found in the earth in a gray compact mass, or else the natural *Borax*, or the salt which is drawn from the water of the river *Nilus*, and many other rivers. And it may be, that all these salts are divers kinds of their *Niter*, but the *Niter* of the moderns is nothing else but Salt-peter, and this is that of which I intend to speak.[20]

Since it was regarded as being of mineral origin, the Egyptian nitre or natron also came to be known as mineral alkali or fossil alkali, whereas the salt from plants came to be called vegetable alkali—but these names only reflected the origin, rather than any particular difference in composition. However, as it became possible to refine them to greater purities, eventually subtle differences were noticed between the two types, as the German chemist Wiegleb describes in 1789:

Hitherto there are but two different fixed alkalies known in Nature, viz. the *mineral*, and the *common vegetable* alkali. The *mineral* alkali is distinguished from the other by a less fiery taste. Neither does it deliquesce when exposed to the air, but only falls to a powder, and consequently loses its own water instead of attracting any from the air.[21]

We would now say that the mineral alkali is mainly sodium carbonate, whereas the vegetable alkali is potassium carbonate. When freshly prepared, potassium carbonate takes in water from the atmosphere and turns from a solid to a liquid (it deliquesces), but sodium carbonate is prepared as crystals containing much water of crystallization: that is, water trapped into the structure of the crystals. This water is lost on standing in the open air, and the beautiful large transparent crystals gradually crumble into powder.

Both the vegetable and mineral alkalis were described as fixed alkalis, because when they were heated strongly a residue was left behind. This was in contrast to the third well-known alkali of the time, the volatile alkali (ammonium carbonate), which seemed to disappear without trace on heating.

Sal Ammoniac—the Salt of Amon

Nitre, or natron, was not the only alkali from Northern Africa—sal ammoniac had also been used since ancient times. Just as with most other salts, the precise compound associated with this name has probably changed over time. It is almost certain that 'sal ammoniac' initially referred to impure common salt, sodium chloride, from a particular region. Later, it came to be associated with one particular compound: ammonium chloride, a salt characterized by its volatility (a gentle heat easily converts it to gases which reform the solid on cooling), and the ease with which it produces the sharp-smelling gas ammonia.

Barba in his *Art of Metals*, first published in Spanish in 1640, describes this salt:

> Among all the Salts that Nature alone produceth the scarcest, but of greatest vertue, is the Salt-Ammoniac; they call it vulgarly Armoniac, and from that name conclude, that it comes from *Armenia*, but that is not the true name of it, but Ammoniac, which in Greek signifies, Salt of the sand: and underneath the sand (of the Sea shore, I suppose) it is found congealed in little pieces by its internal heat, and the continual burning of the Sun, baked so much, that it is made the bitterest to taste of all kind of Salt. Goldsmiths use it more than the Physicians. It is one of those they call the four spirits, because the fire will convert them into smoak, and so they fly away.[22]

In 1797, Stephen Dickson (the Irish critic we encountered in the last chapter) gave his account of how it got its name, dating back to the biblical story of Lot, whose poor wife had earlier been turned into a pillar of salt as the family fled the destruction of Sodom:

> When the younger daughter of Lot had a child by her father, she called his name Ben Hammi. He was the father of the Ammonites (*Genesis*, xix. 38) who inhabited that part of Libya which adjoins the Mediterranean. This territory was called Ammonia…Hence the muriated volatile alkali, which abounded in this country, was called Ammoniacal Salt, or Sal Ammoniac.[23]

The kingdom of Ammon was in what is now Jordan, and the modern capital, Amman, derives from its earlier name Rabbath Ammon, essentially meaning 'capital of Ammon'.

Most authors give a rather different derivation from Dickson. Pliny speaks of 'the salt Ammoniacum, so called, by reason that it is found under the sands', and mentions how 'they practised to dig in the desart & drie sands of Affricke, and found more as they went, even as far as to the temple and Oracle of *Iupiter Ammon*'.[24] The temple referred to, dedicated to the Egyptian god Amun (who later became associated with the Roman god Jupiter), was located at the Siwa Oasis in Ancient Libya, now part of Egypt. Travellers came from far and wide to consult the famous oracle based here, with notable patrons including Perseus (prior to beheading Medusa), Hercules, and in 332 BC Alexander the Great, who the oracle confirmed to be no less than the descendant of the god Amun himself. Charas in his *Royal Pharmacopea, Galenical and Chymical* of 1678 tells us how the salt was formed here:

> THE name of *Ammoniack*, giv'n to this Salt, has carry'd it at all times from above thirty other names which Authors have giv'n it, the repetition whereof is not necessary. The Temple of *Jupiter* Ἀμμον [Ammon], situated in the midst of the Deserts of *Lybia*, gave it its name; because this Salt was formerly found sublimated upon the superficies of the burnt Sands of that Country.
>
> The Urine of Camels that generally travell'd that way in Caravans, in the pilgrimages that were continually made to this Temple, was the first and principal matter, and the acid Salt of the air, which impregnated this Salt in the night time, by its union stopp'd the volatile parts, which the heat of the Sun had otherwise dissipated.[25]

Charas even supplies us with a picture of a camel helping to prepare this salt of Amun—one that was copied towards the end of the eighteenth century in a design for decorating porcelain ware (Figure 39). Irrespective of what the original sal ammoniac may have been, the one described by Charas is certainly ammonium chloride, ultimately composed of ammonia (present in old urine) and hydrochloric acid (the acid formed from sodium chloride salt).

A detailed preparation of ammonium chloride from salt and urine is described in the sixteenth century by Alexis of Piedmont in his *Book of Secrets*:

> To prepare salte armoniacke
>
> Take ten pounde of prepared salt, and powre upon it some warme pisse of a man that is in health, and hath not dronke but wyne, and let the salt dissolve in the sayd pisse, and go to the bottome, then straine it thorow a felt into a caudron, put to it some soute of a bakers oven, boyling it together: When this salte is drye: powre upon it some mans pisse, and do this so long untill the ten pots of urine be consumed in the ten pound of salt.[26]

Charmingly, Alexis warns the reader, 'You muste take heede, that the caudron ronne not over, whan the Uryne boyleth'—wise words indeed. The original use of camel's urine rather than human might have been, as another author puts it, since ''tis not so stinking'.

Heating ammonium chloride, or even just urine, with one of the fixed alkalis (the hydroxide or carbonate of either potassium or sodium) liberates ammonia

CHAMEAU DE L'ASIE.
FAISANT LE SEL AMMONIAC.
Dessinée par J. Charton.

Fig. 39. Based on a figure from the pharmacopoeia of Charas, this engraving showing 'an Asian camel making sal ammoniac' was included in a collection of designs by Jacques Charton published in France in the 1780s.

gas. This was known as 'the volatile spirit of salt ammoniack', or 'spirit of urine', and is described by Glauber in his *Description of New Philosophical Furnaces* from 1651: 'Out of urine or salt Armoniack a powerful and penetrating spirit may be made several wayes, which not only is to be used in physick for many diseases, but is also found very useful in mechanical and Chymical operations.'[27]

Glauber uses the urine 'of sound men living chaste' heated with 'calcined Tartar' (the potassium carbonate we encountered earlier), but thoughtfully adds 'because the spirit of urine is tedious to make, therefore I will shew, how to get it easier out with salt Armoniack'. He later describes the ammonia as a spirit 'of a sharp penetrating essence, and of an airie, moyst and warm nature', and notes among its uses that 'onely smelled unto it cureth the megrim and other Chronical diseases of the head: for it dissolveth the peccant matter & evacuateth it through the nostrils'.

The first person to prepare and isolate pure ammonia gas was Joseph Priestley, who prepared it by heating a mixture of 'one fourth of pounded sal ammoniac [ammonium chloride], with three fourths of slaked lime [calcium hydroxide]'.[28] Since the gas dissolves extremely readily in water, he collected the gas by the displacement of mercury rather than water. Priestley called the gas 'alkaline air', and 'with the same ease I also procured this air from *spirit of hartshorn*, and *sal volatile* [ammonium carbonate] either in a fluid or solid form, i.e. those volatile alkaline salts which are produced by the distillation of sal ammoniac with fixed alkalis'. Spirit of hartshorn was an old preparation obtained by the distillation of the horns of deer. When heated, this animal matter decomposed and produced in part of the fraction aqueous ammonia and ammonium carbonate.

The French Reform of the Alkalis

When Lavoisier and his colleagues presented their revision of chemical nomenclature in the 1780s, they wanted to remove compound descriptive names such as vegetable alkali, mineral alkali, and volatile alkali and replace them with single words. The Swedish mineralogist Torbern Bergman had tried to do this earlier and had suggested *potassinum*, *natrum*, and *ammoniacum*. The French chemists went along with *potasse* (potash in English) and *ammoniaque* (ammoniac), but preferred *soude* (soda) to natrum:

> The word *potash* has been used to signify vegetable fixed alkali obtained by the washing of ashes; we propose only to annex to this expression the idea of purity and causticity.
>
> We have preferred the word *soda* to that of natrum, particularly because it was more universally known; every chymist is acquainted with the words sal sodae or the crystals of soda, and the substance to which we give the name of

soda is precisely that which constitutes the crystals of sal sodae, excepting the carbonic acid which occasions the crystalline form.[29]

The authors actually used these names to refer not to the carbonate alkalis, but the simpler, much more corrosive alkalis that can be prepared from them if the carbon dioxide (carbonic acid) is removed. As Black had shown, carbon dioxide could be driven out of carbonates by the action of heat or acids. While it is extremely difficult to drive out the carbon dioxide from sodium or potassium carbonate using heat alone, this is not the case with chalk or lime (calcium carbonate), which more easily forms solid calcium oxide, known as quicklime. The calcium oxide formed, even after it has cooled down, reacts violently with water in a process known as slaking, generating sufficient heat to turn the water to steam which then hisses out, causing particles of rock to fly off in an animated fashion. This explains the old use of 'quick' in its name, meaning 'living', just as in 'quicksilver'—the living, liquid metal, mercury.

The solid remaining after calcium oxide has reacted with water, calcium hydroxide, was known as slaked lime, and its aqueous solution as lime water. It is this lime water that can be used to prepare the more caustic form of the mineral and vegetable alkalis. When solutions of calcium hydroxide and either potassium or sodium carbonate are mixed, insoluble calcium carbonate (chalk or lime) is precipitated, leaving a solution of either potassium or sodium hydroxide. These alkalis are sold today as caustic potash and caustic soda.

Although the chemistry was not understood, this process had been utilized for many years. In the 1660s, Tachenius described how the alkali could be made during the manufacture of soap, and went on to give a gruesome account of just how caustic this 'fiery alkali' could be:

> Therefore Soap-men add to the *Calx* [calcium oxide] a factitious *Alcaly*, burnt out of Vegetables in a triple proportion, because it moretifies the *Acid* part in the *Calx*, and melts the other part by its like...then with a sufficient quantity of water they extract the Lixivious Fiery *Alcaly*; (I call it Fiery, because this boiling *Lixivium*, or Ley, consumed in a moment a Drunken Man with his Wollen Cloaths, so that nothing of him was found but his Linnen Shirt, and the hardest Bones, as I had the Relation from a Credible Person, Professor of that Trade)...[30]

After the French reform, it was these caustic hydroxides that were meant by the simple terms 'potasse' and 'soda'—clearly an improvement on the more wordy 'caustic vegetable fixed alkali' and 'caustic mineral fixed alkali'. The carbonates themselves were termed in French '*carbonate de potasse* or *carbonate de soude*', giving us the terms will still use today.

Needless to say, Stephen Dickson did not like the new French terminology. His specific objection (and the reason why the Germans preferred to use the word '*kali*') was that 'potash' was a commercial term used for semi-refined vegetable alkali, the further refined substance being known in the trade as 'perl-asche' (or 'pearl-ash') from its improved whiteness. Dickson notes, but rejects, the suggestion from Black and his colleagues of the word 'lixiva', reflecting the process of extracting the alkali from ashes. A similar variant suggested by Wiegleb, 'spodium' (from the Greek for 'ashes'), also never caught on, and neither did 'tartarin' (after 'salt of tartar'), suggested by Irish chemist Richard Kirwan.

Dickson also had things to say about 'soda'. While many believed that the word was derived from Arabic, he thought this unlikely, since 'soda' 'means a pain of the head: and is derived from a word which signifies the temple'.[31] He thought it was more likely to be derived 'from the German *sode*, which is ebullition, and comes from *sieden* to boil and that from *sod* water', since the alkali was obtained by boiling aqueous solutions. He noted it had also been suggested as deriving from the French '*soude*', from its use as a flux in soldering metals. Dickson rejected the term 'natron' (preferred by the Germans) because of the historical confusion with nitre and what that term now came to signify (potassium nitrate), but he felt less strongly about Black's related suggestion 'trona', a commercial name used for the naturally occurring sodium carbonate.

Dickson himself proposed the terms 'plankali' (from 'plant kali'), 'foskali' (from' fossil kali'), and 'volakali' (from 'volatile kali, ammonia') for the alkalis, but in the end it was '*potasse*', '*soude*', and '*ammoniaque*' (and their variants) which were settled on in France and England, but '*kali*', '*natrum*', and '*ammonium*' in Germanic lands. These names of the compounds were to determine the names of the elements that were eventually isolated from them.

The Prediction and Isolation of the Metals

In their system of nomenclature, the French authors proposed names for 'simple substances, or such as have not as yet been decomposed'.[32] They included ammoniac even though one of their authors, Berthollet, had correctly shown ammonia to be 'a combination of azot [nitrogen] and hydrogen'.[33] They suspected the fixed alkalis to be compounds from which new elements would also eventually be isolated, and in a prescient manner correctly named the carbonates and other salts of soda and potassa before the elements sodium and potassium themselves were isolated. They further named salts of lime (calcium), magnesia (magnesium), and barytes (barium) before any of these metals were known. Remarkably, all five of these extremely reactive metals were first isolated by one man: the young Cornish chemist Humphry Davy.

The Birth of Electrolysis

Although static electricity had been studied for many years, the first production of a steady electrical current was achieved around 1800 by the Italian scientist Alessandro Volta (1745–1827), who essentially made the first battery by stacking together discs of silver and zinc separated by cloth soaked in a solution of salt. The inspiration for this invention came from the earlier observation of his fellow Italian Luigi Galvani, who noticed how dissected frogs' legs twitched when they came into simultaneous contact with two different metals. Volta announced his discovery with a paper read before the Royal Society of London, and after being shown the initial correspondence, the surgeon Anthony Carlisle immediately set about constructing his own 'voltaic pile' using seventeen silver half-crown coins and matching discs of zinc. On 30 April 1800, Carlisle and his friend, scientist William Nicholson, began to experiment.

After first using their battery to shock themselves in the arm, they set to more serious investigations, noting that the current was transmitted through the usual conductors of electricity, but not through non-conductors such as glass. In order to make sure that the contacts to the pile were good, they added a drop of water to the uppermost plate and 'Mr Carlisle observed a disengagement of gas round the touching wire.'[34] Remarkably, Nicholson noticed that this gas, 'though very minute in quantity, evidently seemed to me to have the smell afforded by hydrogen when the wire of communication was steel'. This prompted them 'to break the circuit by the substitution of a tube of water between the wires'. The tube was filled with 'New river water'. More bubbles ensued, and when, a few days later, they tried placing two flattened wires of the inert metal platinum in the circuit, they noticed different reactions taking place at the two poles of the pile: 'the silver side gave a plentiful stream of bubbles, and the zinc side also a stream less plentiful'. They add: 'It was natural to conjecture, that the larger stream from the silver side was hydrogen, and the smaller oxigen.' Further experiments confirmed this hypothesis, and also showed that exactly twice as much hydrogen was formed as oxygen.

The two gentlemen had made the remarkable discovery that not only could water be synthesized directly from its elements when mixed in the correct proportions, as Cavendish and Lavoisier had shown, but it could also be broken down into them again using electricity. This discovery opened a new era in what came to be called 'electrolysis'—'splitting up using electricity'.

The Basis of Potash Isolated

Humphry Davy immediately started experimenting with his own voltaic piles, presenting a paper in June 1801 and a more extensive review in 1806 detailing experiments to date, including the formation of metals such as iron, zinc, and tin from solutions of their salts in water. But Davy's most important discovery in this area took place on 6 October 1807, when, after failing to produce anything other than the splitting of the water when applying his electric pile to aqueous solutions of the alkalis, he tried with pure molten potash (potassium hydroxide) heated on a platinum spoon. He was using the massive battery he had assembled at the Royal Institution in London, with '24 plates of copper and zinc of 12 inches square, 100 plates of 6 inches, and 150 of 4 inches square'[35] arranged horizontally in troughs to contain the liquid, rather than in vertical piles. After some tinkering with the conditions, 'a vivid action was soon observed to take place'. He writes in his paper announcing the discovery:

> The potash began to fuse at both its points of electrization. There was a violent effervescence at the upper surface; at the lower, or negative surface, there was no liberation of elastic fluid; but small globules having a high metallic lustre, and being precisely similar in visible characters to quicksilver, appeared, some of which burnt with explosion and bright flame, as soon as they were formed, and others remained, and were merely tarnished, and finally covered by a white film which formed on their surfaces.
>
> These globules, numerous experiments soon shewed to be the substance I was in search of, and a peculiar inflammable principle the basis of potash.[36]

Davy's paper does not convey his full excitement at this amazing discovery. During the experiment he was assisted by his younger cousin Edmund, who later reported that 'when he saw the minute globules of potassium burst through the crust of potash, and take fire as they entered the atmosphere, he could not contain his joy—he actually bounded about the room in extatic delight; and that some little time was required for him to compose himself sufficiently to continue the experiment'.[37]

Within a few days, Davy had also, for the very first time, isolated the metal from soda (sodium hydroxide): 'Soda, when acted upon in the same manner as potash, exhibited an analogous result; but the decomposition demanded greater intensity of action in the batteries.'[38]

Most remarkable was the action of these new elements with water—an experiment which still delights students today. During the reaction, the potassium metal darts around on the surface of the water, giving out hydrogen gas which helps produce the beautiful purple flame, and eventually all that is left behind is a solution of potassium hydroxide. Larger pieces react more vigorously, as Davy describes:

> The action of the basis of potash on water exposed to the atmosphere is connected with some beautiful phenomena. When it is thrown upon water, or when it is brought into contact with a drop of water at common temperatures, it decomposes it with great violence, an instantaneous explosion is produced with brilliant flame, and a solution of pure potash is the result.[39]

As these metals are so different from any others so far discovered, Davy initially wonders if they should be called metals at all, but correctly concludes they should, since, 'They agree with metals in opacity, lustre, malleability, conducting powers as to heat and electricity, and in their qualities of chemical combination.'[40]

At last, Davy turns to naming his new discoveries. He notes that the recent trend (one still continued to this day) was that the names for metallic elements should end with the Latinized '-um' or '-ium':

> In naming the bases of potash and soda, it will be proper to adopt the termination which, by common consent, has been applied to other newly discovered metals, and which, though originally Latin, is now naturalized in our language.
> Potassium and Sodium are the names by which I have ventured to call the two new substances: and whatever changes of theory, with regard to the composition of bodies, may hereafter take place, these terms can scarcely express an error; for they may be considered as implying simply the metals produced from potash and soda.[41]

Perhaps Davy was still haunted by the ghost of 'phosoxygen', his element name suggested with the enthusiasm of youth but later retracted. This time, he was more cautious:

> I have consulted with many of the most eminent scientific persons in this country, upon the methods of derivation, and the one I have adopted has been the one most generally approved. It is perhaps more significant than elegant. But it was not possible to found names upon specific properties not common to both; and though a name for the basis of soda might have been borrowed from the Greek, yet an analogous one could not have been applied to that of potash, for the ancients do not seem to have distinguished between the two alkalies.

A translation of Davy's paper appeared in German the following year, but throughout the paper, the translator Ludwig Gilbert (who also edited the journal) uses the terms more common in German, '*kali*' and '*natron*', for the alkalis potash and soda and refers to the new elements isolated from them as the 'basis of Kali' and the 'basis of Natron'. After the section where Davy names his new metals, Gilbert suggests the names *Kalium* and *Natronium* be used in German.[42]

The German names became relevant because Jöns Jacob Berzelius used these versions when he developed the system of symbols that would become the international language of chemistry. He wrote in 1811:

> The French and the British still retain the names of *potassa* and *soda* for these alkalis, because they agree with their language better than *kali* and *natron*. However, as potash indicates a mass containing vegetable alkali, which is an object of trade, we need a different name for this alkali in its state of purity. I used one of kali, adopted by the most distinguished German chemists. It is the same with the names *natrum* and *soda*. Thus we must name the metal bases of the alkali *kalium* and *natrium*, instead of *potassium* and *sodium*.[43]

Early Chemical Symbols

The use of symbols in chemistry dates back to the very earliest alchemical texts. As we saw in Chapter 1, the metals were assigned symbols, but so too were some compounds. When Lavoisier and colleagues published their revised nomenclature, a system of symbols based on the reform was appended to the work. This was devised by two more junior associates: Jean-Henri Hassenfratz, who was an assistant in Lavoisier's laboratory at the time, and Pierre-Auguste Adet, for whom chemistry was more of an interesting pastime and who later became one of the first ambassadors to the newly formed United States. While their symbols look horrifying to modern chemists, they represent progress: there is a level of standardization in the symbols used for the elements, and, perhaps more importantly, the symbols for compounds were to be made up from the symbols of their constituent elements. As the authors put it: 'as the compound bodies are all formed by the different combinations of simple sub-stances, the characters to express compound bodies should be made by the junction of the different characters of simple bodies'.[44]

The most important aim was that the new symbols should be readily under-stood. They write, 'In our reformation of the chymical characters we are far from having the same design with the ancient chymists. They endeavored by every means to screen their science with a mysterious veil from the eyes of the vulgar; we ought on the contrary to use our utmost endeavors to render our knowledge as communicative as possible.'

Previous attempts had been made to bring some uniformity to the archaic symbols. Perhaps the most notable was by Bergman, who, in addition to using the historical symbols for the metals, also proposed symbols to represent, for example, sulfate, nitrate, and chloride (using our modern terms). These could then be combined with a cross (signifying acidity) to represent particular acids, or with the symbols for alkalis or metals to form many different salts. Bergman even went further, giving many examples of chemical equations; an example is shown in Figure 40, together with a modern interpretation. Bergman gave examples both for aqueous solutions and for reactions that take place when the dry solids are heated together.

Despite the significant advances made in Bergman's system, in the eyes of Hassenfratz and Adet, it was far from ideal and at times quite illogical. In their proposed scheme, they divide the 'simple bodies' into different categories gen-erally following Lavoisier's system, such as the elements that commonly make up many different substances (light, caloric, oxygen, and nitrogen), alkaline substances (including soda, potash, and lime), combustible substances (includ-ing carbon, sulfur, and phosphorus), and metals. They also gave characters to substances that had not yet been broken down, such as the different organic acids that could form many different salts.

The common elements making up the first group were represented using either a straight line, drawn at different angles, or in the case of light, a vertical

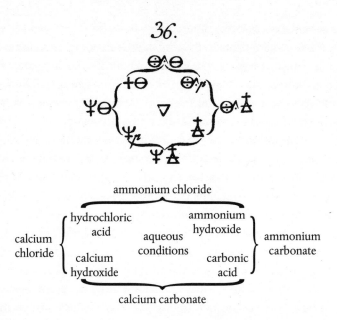

Fig. 40. One of Bergman's chemical equations in symbols from 1775 together with a modern interpretation in words below. What Bergman had summarized is that when an aqueous solution of calcium chloride is mixed with an aqueous solution of ammonium carbonate (the reactants at the sides of the scheme), a precipitate of insoluble calcium carbonate is formed (shown at the bottom), leaving a solution of ammonium chloride (shown at the top). We could also view this as saying that when the ions of chloride, ammonium, calcium, and carbonate are mixed (essentially what is in Bergman's central square), insoluble calcium carbonate is precipitated, leaving the ammonium and chloride ions in solution.

wavy line. Wavy lines at different angles were reserved in case more elements in this category were later discovered. These lines could easily be added to other symbols to form compounds.

For the alkaline substances, they used triangles, pointing upwards for potash and soda and downwards for the alkaline earths, including the oxides of calcium and barium. To distinguish between species within the same class, they proposed 'inscribing in the triangle, which expresses each species of alkali or earth, the first letter of the Latin name of that particular substance'.[45]

They kept the alchemical symbol for gold, the circle with a dot in the centre, 'merely for the sake of preserving the ancient character'. For the other metals, they retained the circle but now inserted the initial letter of the Latin name of each metallic substance. They 'preferred the Latin initial letters, because the Latin names are universally known'. Occasionally, the Latin names of two elements would have the same initial letter. In this case, one would be represented by the single letter, for the other 'the initial letter of the second substance united to the consonant next in order'. They give an example to clarify; 'silver, whose Latin name begins with an A, like arsenic, is represented by a circle in which is inscribed the letter A, while the sign of arsenic is a circle containing an A and an S joined together'.[46]

The use of Latin was an important feature, since Hassenfratz and Adet note 'how much it is necessary to have characters of chymistry common to all chymists' so that even scientists from different countries would easily be able to communicate and understand one another using exactly the same characters. Unfortunately, while attempting to make things clearer for their own countrymen, some translators chose to modify the scheme.

Although George Pearson did not include Hassenfratz and Adet's characters in the first edition of his *Translation of the Table of Chemical Nomenclature*, published in 1794, he did in his expanded second edition of 1799. He decided to use the English names for the symbols, and was severely criticized for doing so by Richard Chenevix in his *Remarks Upon Chemical Nomenclature, According to the Principles of the French Neologists*, published in 1802. Chevenix states:

> Dr Pearson has committed a radical fault in using the initials of the English names, instead of those of the Latin. By this he has circumscribed the limits of the language, and, from a universal character, reduced it to a provincial dialect. The learned world may be considered as forming an empire, of which England, France, Germany, Italy, America, &c. are provinces. If these nations all assume to themselves the right of speaking their own language, they will soon cease to understand one another.[47]

Not only did English authors use different symbols from Hassenfratz and Adet, but to add to the confusion, different authors or translators used different symbols. Some examples of the letters inscribed within circles from a number of English texts are given in Table 1.

A circle with an S inside could have meant tin (from the Latin 'stannum'), silver, or sulfur; a circle with a C could have meant copper (from the Latin 'cuprum'), cobalt, or carbon; a circle with an A could have meant silver (from the Latin 'argentum'), antimony, or azote (nitrogen). Furthermore, the symbol for the alkali potash varied between an upwards-pointing triangle with either a P (for potash), V (for vegetable alkali), or L (for lixia), while for soda it was either S (soda), F (fossil alkali), or T (trona).

Table 1. Some different symbols which were inscribed in circles and used to represent a selection of elements. All the works were published in English with the names of the authors indicated.

Element	St John	Pearson	Bouillon-Lagrange	Kerr	Duncan
	1788	1799	1800	1802	1803
Mercury	H	Q		Me	H
Copper	C	Co	C	Cu	Cp
Cobalt	K	C		Co	Cb
Carbon				C	
Silver	A	Si	S	Ar	Ag
Antimony	ST	A	A	An	Sb
Tin	S	Ti	T	St	Sn
Strontium		St			
Sulfur				S	
Azote				A	

The symbols were included in a number of different works, but they were always present as complete tables rather than actually being used as originally intended within the text of the book itself to indicate the compounds and reactions being described. In 1800, the translator of Gren's *Principles of Modern Chemistry* (thought to be a Dr Gruber) added such a table as an appendix, using Hassenfratz and Adet's original letters, but nonetheless noted:

> It must be confessed that the contrivance is very ingenious; but it does not promise to be of great utility.—For if employed in the text of a printed book, it deranges the lines, and increases the expences of printing and waste of paper, by the interstices which it causes between the lines, even if types should be cast for them. Besides, whether written or printed, it is easier for any person to read, for instance, *muriat of ammoniac*, than to strain his eyes, or his attention, to avoid mistaking one sign for another; and in the same manner it requires less trouble in writing muriat of ammoniac, than to trace its corresponding compounded characters neatly and distinctly enough to prevent the reader's falling into error.[48]

He later adds, '[I]f it is no longer intended to deal in scientific secrets, in the manner of alchemists, is there any more occasion for such characters?'

The Modern Symbols

One person who did use the symbols of Hassenfratz and Adet, at least in his manuscripts, was Berzelius. Despite using them, Berzelius was unhappy with the symbols. By 1813 he published the first suggestion of a new system, which he expanded the following year under the heading *On the Chemical Signs, and the Method of Employing Them to Express Chemical Proportions.* He writes:

> When we endeavour to express chemical proportions, we find the necessity of chemical signs. Chemistry has always possessed them, though hitherto they have been of very little utility. They owed their origin, no doubt, to the mysterious relation supposed by the alchymists to exist between the metals and the planets, and to the desire which they had of expressing themselves in a manner incomprehensible to the public. The fellow-labourers in the anti-phlogistic revolution published new signs founded on a reasonable principle, the object of which was, that the signs, like the new names, should be definitions of the compositions of the substances, and that they should be more easily written than the names of the substances themselves. But, though we must acknowledge that these signs were very well contrived, and very ingenious, they were of no use; because it is easier to write an abbreviated word than to draw a figure, which has but little analogy with letters, and which, to be legible, must be made of a larger size than our ordinary writing. In proposing new chemical signs, I shall endeavour to avoid the inconveniences which rendered the old ones of little utility.[49]

His proposed solution to avoid the difficult-to-reproduce symbols was to simply use letters (without the surrounding shapes employed by Hassenfratz and Adet):

> The chemical signs ought to be letters, for the greater facility of writing, and not to disfigure a printed book. Though this last circumstance may not appear of any great importance, it ought to be avoided whenever it can be done. I shall take, therefore, for the chemical sign, the *initial letter of the Latin name of each elementary substance*: but as several have the same initial letter, I shall distinguish them in the following manner: 1. In the class which I call *metalloids*, I shall employ the initial letter only, even when this letter is common to the metalloid and to some metal. 2. In the class of metals, I shall distinguish those that have the same initials with another metal, or a metalloid, by writing the first two letters of the word. 3. If the first two letters be common to two metals, I shall, in that case, add to the initial letter the first consonant which they have not in common: for example, S = sulphur, Si = silicium, St = stibium (antimony), Sn = stannum (tin),

C = carbonicum, Co = cobaltum (cobalt), Cu = cuprum (copper), O = oxygen, Os = osmium, &c.[50]

This paper was first published in English, but Berzelius's manuscript had been translated into this language by the editor of the journal, Thomas Thomson—yet another Scottish chemist who had been inspired to take up the subject by the brilliant Joseph Black. However, Thomson committed the terrible sin of changing some of the element names, and worse still, the symbols, in order to make the essay 'intelligible to the English reader'. Among other changes, for Davy's metal sodium he used the symbol 'So' and for potassium he used 'Po'. Berzelius was furious. A translation of his letter (in French) to Thomson reads:

> You have given yourself license in some places to change the Latin nomenclature that I have used; I consider it the right of the author to use such nomenclature he has chosen. You violated that right and I ask you to indicate to readers of your Annals that you did it without my consent...[51]

With regard to Thomson's heretical changes of the names and symbols for sodium and potassium, Berzelius referred to the rationale mentioned in his 1811 essay on nomenclature (see earlier), and continued:

> we might add further to this the absurd derivation of a Latin word for 'pot' and 'ash', both Gothic words, preserved as much in English and German as they are in Swedish.

Thomson wrote back to Berzelius:

> The chemical nomenclature in the English language is too well established to be altered either by your opinion or by mine and if we wish to be read we must conform ourselves to it even though in our opinion the words might be improved. Beryllia, Kalium, Natrium, would not have been understood. Dr BLACK tried to bring into use Trona and Lixiva for Soda and Potash, but his attempt failed. KIRWAN proposed Tartarin with as little success. So would mine were I to propose Kalium and Natrium. In Germany they were used because KLAPROTH had already brought Kali and Natron into general use. Here we are not so fond of changes.[52]

Thomson was not alone in deviating from Berzelius's system. The French mineralogist Beudant also used his own version of symbols, including So and Po

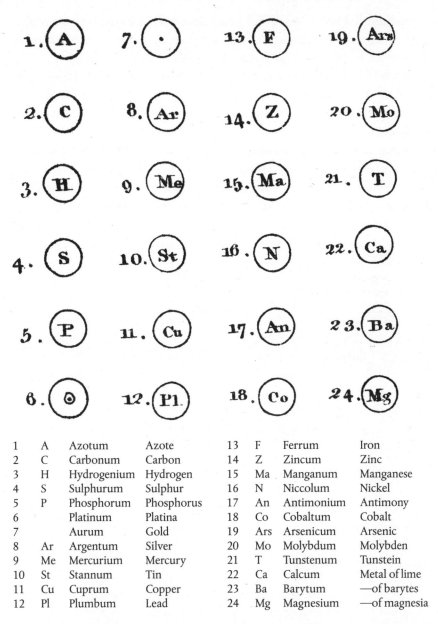

1	A	Azotum	Azote	13	F	Ferrum	Iron
2	C	Carbonum	Carbon	14	Z	Zincum	Zinc
3	H	Hydrogenium	Hydrogen	15	Ma	Manganum	Manganese
4	S	Sulphurum	Sulphur	16	N	Niccolum	Nickel
5	P	Phosphorum	Phosphorus	17	An	Antimonium	Antimony
6		Platinum	Platina	18	Co	Cobaltum	Cobalt
7		Aurum	Gold	19	Ars	Arsenicum	Arsenic
8	Ar	Argentum	Silver	20	Mo	Molybdum	Molybden
9	Me	Mercurium	Mercury	21	T	Tunstenum	Tunstein
10	St	Stannum	Tin	22	Ca	Calcum	Metal of lime
11	Cu	Cuprum	Copper	23	Ba	Barytum	—of barytes
12	Pl	Plumbum	Lead	24	Mg	Magnesium	—of magnesia

Fig. 41. The symbols used in 1802 by Kerr in his translation of the work of Adet and Hassenfratz. Ten of the symbols now used for the elements are the same as Kerr's (minus the circle). Symbols appear for calcium, barium, and magnesium, but these metals had not actually been isolated at this time.

for sodium and potassium, O for gold (Or), and Ox for (Oxigène) in the first edition of his textbook published in 1824, but after criticism from Berzelius, he conformed to the Swedish system in the second edition.

Despite some interesting variations along the way (including some bizarre modifications from Berzelius himself, which involved substituting the characters for some elements with dots and commas), over time, a common system of symbols evolved.

In many ways, the elegant system invented by Berzelius was a logical progression from the system of Hassenfratz and Adet. What is remarkable (and not entirely coincidental) is that ten of the symbols used by the English translator Kerr in his interpretation of their system are still in use today, whereas just two of the Frenchmen's original system have survived unchanged. Part of the reason for this is because Kerr used symbols (inscribed in circles) for carbon, hydrogen, sulfur, and phosphorus, which he thought ought to be treated in an equal manner to the metals, since, for example, they could all be made to combine with oxygen. But the other reason for his high success rate was that he thought the elements calcium, barium, and magnesium had been isolated back in 1791, and gave them their modern symbols in 1802 (Figure 41). As we shall see in the next chapter, on this point he was quite wrong, and it was not until 1808 that all three were isolated by the brilliant Humphry Davy. But even Davy did not end up with all his element names used in the form he suggested. In order to understand why, we will need to take a closer look at these elements that make up Group 2 of the modern periodic table, a group also known as the alkaline earths.

6

LOADSTONES AND EARTHS

Since a line must be drawn between salts and earths,
I think it should begin where solution is scarcely perceptible.

—Kirwan, 1794[1]

The Five Earths

Jöns Jacob Berzelius (1779–1848), discoverer of the elements selenium, thorium, cerium, and silicon and deviser of the chemical symbols we use today, was one of the last in a long list of Swedish mineralogists and chemists active during the eighteenth century. Berzelius himself regarded one of his predecessors, Axel Fredrik Cronstedt (1722–65), as the founder of chemical mineralogy. We met Cronstedt in Chapter 2 as the discoverer of the element nickel, isolated from the ore *kupfernickel*. But another of Cronstedt's achievements was perhaps of even greater significance: his development of a classification of minerals based not on their physical appearances, as had been common up to this time, but on their chemical compositions. He first published his scheme anonymously in Swedish in 1758, but it was later translated into English as *An Essay towards a System of Mineralogy*. Cronstedt recognized four general classes of minerals: earths, bitumens, salts, and metals. As their name suggests, the bitumens were flammable substances that might dissolve in oil but not in water. The main difference between the salts and the earths was that the former, which included the 'alcaline mineral salt' natron, could be dissolved in water and recrystallized from it. The earths he defined as 'those substances which are not ductile, are mostly indissoluble in water or oil, and preserve their constitution in a strong heat'.[2]

Cronstedt initially recognized nine different classes of earth. By the time of Torbern Bergman (1735–84), these had been reduced to five which 'cannot be derived from each other or from anything simpler'.[3] Lavoisier and his collaborators included these five in their great work on nomenclature even though they suspected that, like soda and potash, they were most likely not simple substances, but species that contained new metals. In the 1788 English translation of the nomenclature these were called silice, alumina, barytes, lime, and magnesia. The first two eventually, in the early nineteenth century, yielded the elements silicon and aluminium. The word 'silicon' derives from the Latin *'silex'* (meaning 'flint'—a form of silicon dioxide), with the ending '-on' reflecting its resemblance to the other non-metals carbon and boron. We shall return to the naming of aluminium later, but for now, simply note that it ultimately derives from *'alumen'*, a term for a bitter salt mentioned in Pliny. The name alum was initially used for a number of different substances, but eventually it came to be associated with just one compound: potassium aluminium sulfate. In his 'Essay Explaining Metallick Words', appended to his 1683 English translation of Ercker's classic on mining, John Pettus derives the Latin word *'alumen'* 'from Lumen, in respect of its transparency and nearness to Christal, and is accounted among the brighter stones'.[4]

The other three earths, barytes, lime, and magnesia, were noted as having rather similar properties: they all effervesced with acid, giving out carbon dioxide, and after being heated their residues gave alkaline solutions in water. For this reason they were known as the alkaline earths. In the third English edition of Lavoisier's *Traité*, the translator Kerr added to these three the newly discovered strontites (strontium carbonate), named after the parish of Strontian—an area in western Lochaber, Scotland, where the mineral was discovered in a lead mine. Apparently the Gaelic name of the parish, *Sròn an tSìthein*, literally means 'nose of the fairy hill', referring to a prominent hill in the region.[5] A couple of years later, a new earth was discovered in the precious stone beryl. This was initially named glucine, from the Latin for 'sweet', owing to the sugary taste of its (very poisonous) compounds; but the authoritative German mineralogist and discoverer of uranium, Martin Heinrich Klaproth, took exception to this name because the salts of another newly discovered earth called yttria (actually a complex mix of many new earths, which were to be gradually separated in the nineteenth century) were also sweet-tasting. He writes:

> However well chosen the name *glucine* appeared at first, it would in my opinion be better to substitute to it that of *beryllina*, which name would more distinguish and characterize it; since in the *yttria* we are now possessed of another earth, which likewise gives a sweet taste to the neutral salts formed by it.[6]

Although the name glucinium was used for some time, the metal later isolated from the earth eventually became known as beryllium, the lightest member of the Group 2 elements of the periodic table. The complete group, still also known as the alkaline earth metals, are beryllium, magnesium, calcium, strontium, barium, and radium—the last being the intensely radioactive element discovered by Pierre and Marie Curie in 1898. Remarkably, the three oldest known earths of these elements were first broken down and named by Berzelius's great rival, Humphry Davy. While the origins of the names of most elements are clear, it was the name for magnesium that was to cause a headache for Davy because of its complicated history, so intertwined with that of one of nature's most magical gifts, the magnet.

Different Magnets

The Stone which *Magnes* Greeks doe call,
A Stone most wondrous above all;
Which Iron drawes, and that is much,
This Iron drawes other with a touch,
As Loadstones doe—[7]

In the medieval lapidaries, or treatises on stones, many substances are credited with having magical powers or virtues such as enabling the bearer to become indomitable, invincible, or even invisible. Disappointingly for those who still subscribe to the healing power of crystals, evidence for any such properties is lacking. But there is one stone that really does seem to possess supernatural properties, a stone able to move metal from a distance without touching it: the magnet.

Although earlier accounts of magnets were given, notably in China, the descriptions provided by Pliny the Elder in his *Naturalis Historia* helped to contribute to the general confusion later surrounding this stone. When speaking of the loadstone, a naturally occurring magnetic oxide of iron in the form of the mineral magnetite, he says 'yron is the onely mettall which receiveth strength from that stone', and adds that if the iron is 'once well touched and rubbed withall, it is able to take hold of other peeces of yron: and thus otherwhiles we may see a number of rings hanging together in manner of a chaine, notwithstanding they bee not linked and enclosed one within another'.[8] This spectacle would have appeared to be true sorcery, and Pliny adds, 'The ignorant people seeing these rings thus rubbed with the load stone, and cleaving one to another, call it Quick-yron' (yet another example of 'quick' meaning 'living', as we have seen with quicksilver and quicklime). Pliny also relates how the magnet got its name, at least according to the Greek poet Nicander of Colophon, who flourished in the second century BC. It was allegedly named after Magnes, a neatherd (cowherd) on Mount Ida, who, 'as he kept his beasts upon the foresaid mountain, might perceive as he went up and downe, both the hob-nailes which were in his shoes, and also the yron picke or graine of his staffe, to sticke unto the said stone'.[9]

The story was borrowed many times, for example by Samuel Ward in his rather bizarre book from 1640, *The Wonders of the Load-stone or, The Load-Stone Newly Reduc't into a Divine and Morall Use*, in which he draws rather contrived similes between Christ and a magnet. Ward tells us that the name 'load-stone', 'which is peculiar to the English and Dutch, was impos'd upon it, by reason of Leading, directing, and shewing the way' (just like the Saviour).[10] He adds: 'Among names of the second Ranke, it was also called Magnes, because of the great force and virtue of it.'

Far more probable, though, is the idea that the magnet was named after the region in which it was found. The problem is, Pliny describes two distinct regions called Magnesia, and in total, five sundry types of magnets from various localities, all with different properties. He says: 'The principall difference observed in these stones, consisteth in the sex (for some be male, others female;) the next lieth in the colour.'[11] He says that some are red and some are black. That from the Troad, a region in the north-western part of Anatolia, Turkey, 'is blacke, and of the female sex, in which regard it is not of the virtue that others be'—that is, it is a weaker variety. Worse still is the magnet from Magnesia in Asia, which is white and resembles pumice stone and has no attraction for iron—it seems strange that this should be called a magnet at all, and it is likely to have been a completely different mineral, as we shall see.

While Pliny's account of the magnet's attraction to iron is quite accurate, there were also more outlandish powers reputedly associated with the magnet. In his book on minerals, the thirteenth-century German saint Albertus Magnus reports earlier accounts of uses of the loadstone, such as testing for the infidelity of one's spouse: 'If thou wilt knowe whyther thy wife is chaste, or no. Take the stone, which is called Magnes in English, the lode stone . . . Laye thys stone under the head of a wyfe, & yf she be chast, she wil embrace her husbande, if she be not chaste, she wil fall anone forth of the bedde.'[12] The same story also featured in Latin verse in the lapidary of the eleventh-century Bishop of Rennes, Marbode. An English translation of this verse, from 1658, is shown in Figure 42, together with a more sensitive alternative translation from 1860.

If one would know her leads a whorish life,
Under her head, when that she sleeps, it shows:
For she that's chast, will presently imbrace
Her husband whilst she sleepeth; but a whore
Falls out o'th'bed, as thrown out with disgrace,
With stink o'th'Stone, which shows this, and much more.
i

For should'st thou doubt thy wife's fidelity
Unto her slumbering head this test apply;
If chaste she'll seek they arms, in sleep profound
Though plung'd: – th'adultress tumbles on the ground:
Hurled from the couch, so strong the potent fume,
Proof of her guilt, diffused throughout the room.
ii

Fig. 42. A translation from 1658 of Marbode's verse on the loadstone from his eleventh-century lapidary (above) together with a more sensitive alternative translation provided in 1860 by Rev. C. W. King (below).

In a similar manner, Albertus reports how, using the loadstone, thieves may usefully clear a house of its sleeping inhabitants prior to setting about their nefarious business: 'yf thys stone be put brayed, and scattered upon coles, in foure corners of the house, they that be slepynge, shall flee the house, and leave all'.[13]

He also repeats the story, popular in the Middle Ages and falsely attributed to Aristotle, of how magnetic mountains could destroy passing ships by sucking all the iron nails out. This is picked up and dramatically illustrated in the *Grete Herball* from 1526 and other editions of the *Hortus Sanitatis* (Figure 43).

While the idea of magnetic mountains destroying ships is clear fantasy, there is in fact a mountain, Kediet ej Jill in Mauritania, made solely of magnetite. It is located on the western side of Mauritania, about 50 km from the border with Western Sahara, and is clearly visible on satellite maps as a dark 'beauty spot' on the pale, barren Sahara. Not being on the coast, it could not have damaged any ships, and it is unable to extract nails anyway; but it is notable that compasses do not function on the mountain.

As well as describing the fantastic properties of the magnet, Albertus gives accounts (from Aristotle) of magnets that attract things other than iron: 'for some attract gold, and others, different from these, attract silver, and some tin, some iron, and some lead'.[14] Dorothy Wyckoff, the scholar who prepared this translation in the 1960s, suggested that rather than being stones that physically attract the different metals, these accounts may instead be referring to minerals

Fig. 43. A woodcut from the *Ortus sanitates*, published in the late fifteenth century. As a ship sails past a magnetic mountain, the iron nails are drawn out of it, thereby destroying it.

used in metallurgical processes that have a strong affinity for the specified metal. For example, in Chapter 2 we saw how antimony could be used to 'attract' base metals and leave refined gold in the process of cupellation. However, Albertus goes on to list further kinds of ever more fanciful magnets which become harder to rationalize, including varieties that attract human flesh ('it is said that a man attracted by such a magnet laughs, and remains where he is until he dies, if the stone is very large'), bones, hairs, water, and fish. He even claims that 'there is a magnet called "oily" that attracts oil, and a "vinegar stone" that attracts vinegar; and a "wine stone" that attracts wine'. The sixteenth-century Italian metallurgist Vannoccio Biringuccio (whom we encountered in the previous chapter when looking at saltpetre) teases Albertus, writing: 'thus there is lacking only the one [magnet] that produces greens and salt for men, so that, possessing it, they could make a salad where they might be, and having a plate and a little bread, they could have a fine meal!'[15]

It is yet another type of magnet, mentioned by Pliny, that concerns us here. After describing the discovery of glass by heating nitre on the beach, Pliny adds, 'But afterwards (as mans wit is very inventive) men were not content to mix nitre with this sand, but began to put the Load-stone [*magnes lapis*] among, for that it is thought naturally to draw the liquor of glasse unto it, as well as yron.'[16] It is not clear what he means by this particular *magnes lapis*. Perhaps it could have been some sort of calcium-containing limestone necessary to keep the glass stable. Whatever the initial reason, it soon became commonplace for glassmakers to routinely add 'load-stone' to the mix when making glass.

Manganese: The Glassmaker's Magnet

The addition of certain minerals to molten glass was soon found to give rise to a variety of different colours; for example, the addition of cobalt ore produces a beautiful blue colour. Other additives could remove the green colour of crude glass (caused by traces of iron) and produce a more highly valued clear glass. Van Helmont, the inventor of the word 'gas', wrote: 'Also by another Phantasie, doth the Load-stone draw any thing out of Glasse throughly boyled or melted by Fire; for a very small Fragment thereof, being cast into a Mass or good quantity of Glass, while it is in boyling, of Green, or Yellow, makes it White.'[17] He adds that the magnet 'attracteth and consumeth the tinged Liquor out of the Fiery Glass'. However, the substance necessary for this process is not the naturally occurring magnet magnetite (an oxide of iron), but a compound easily confused with this mineral because of its similar appearance, now known to be an oxide of the element manganese.

It was recognized in the Middle Ages that the mineral used to 'cleanse' glass was not the same as the usual loadstone. In his lapidary, after the entry on the loadstone titled (in Latin) 'Magnes' or 'Magnetes', St Albertus Magnus included a separate entry called 'Magnesia' or 'Magnosia' for this stone used by the glassmakers. Biringuccio, writing in Italian in 1540, notes its ability not only to cleanse glass when present in small quantities but also to colour it a beautiful violet when present in larger amounts, and he refers to the mineral using the name *manganese*. This word was also used by the Florentine priest Father Antonio Neri, who published the definitive work on glassmaking, *L'Art Vetraria*, in 1612. In an English translation of this work, *The Art of Glass*, which appeared fifty years later, and in many subsequent works based on it, the name manganese was retained for the mineral.

Occasionally, other names were used for the glass-whitening mineral, still based on its key property. For example, in Macquer's chemical dictionary of 1777 we find under the entry 'Vitrification', when referring to the colours of glass, 'These colours are destroyed by *manganese*, which being added in small quantities, clears the glass, and is therefore called by artists *the soap of glass*.'[18]

Later in the nineteenth century, Austrian mineralogist Wilhelm Karl Ritter von Haidinger coined the term now most commonly used for this mineral—'pyrolusite', derived from the Greek words for 'fire' and 'the action of washing', since, when heated, it 'washes' the glass.

The ancient connection of the name manganese to the loadstone was preserved in the glass-making manuals. In one from 1699 we find: 'Lucretius would perswade us that the Name of Magnes was given to the Load-stone from Magnesia, a certain Country in Lydia, near Macedonia, where it is found, so it is no wonder that that Species of it we use in Glass retains the Name of Magnese and so Manganese, since the Country called by that Name produces it.'[19]

The problem was that as the term 'manganese' became adopted in English for the mineral pyrolusite, the use of 'magnesia' was still common in other languages, and was especially prevalent in Latin texts. For example, while in the 1770 English translation of Cronstedt's classic work on mineralogy the word 'manganese' is used, in the original Swedish version he refers to the ore as *magnesia* or *brunsten* (literally 'brown-stone') and simply notes its other names, including *mangonese* in French. Like many others of the time, Cronstedt is unsure what brunsten (pyrolusite) actually is. While most think it is some sort of iron mineral, he thinks it contains very little metal and is actually a type of earth. It was his fellow countryman Scheele who was the first to study the mineral thoroughly, during which remarkable investigation he also discovered another new element that was present as an impurity, and the poisonous gas chlorine. But even the great Scheele was unable to isolate a metal from brunsten ore. Although an Austrian chemist, Ignatius Gottfried Kaim, claimed (somewhat improbably) to have isolated a number of new metals including manganese in 1770, there is no doubt that metallic manganese was prepared in 1774 by Johan Gottlieb Gahn, the young laboratory assistant of Torbern Bergman—who also famously introduced the hitherto unknown apothecary Scheele to his professor, thereby bringing this superb experimentalist and his work to the attention of the world's scientific community. Bergman reported Gahn's discovery, writing: 'The mineral substance which is called black, or glass-makers, magnesia, is scarcely any thing more than the calx of a new metal.'[20] Bergman says he suspected the mineral might contain a metal but was unable to extract any. However, 'at length, Mr Gahn, without knowing any thing of my experiments, succeeded in obtaining larger pieces of regulus by means of a most intense heat'. The term 'regulus' (the 'king') refers to the metal of a particular ore, and Bergman goes on to give all the details necessary to prepare the 'regulus of manganese'. Once again, though, while the English translator called the new metal manganese (the term we still use today), Bergman in the original Latin has a different name for it. Confusingly, he writes, 'it is called *Magnesium*, to avoid excess verbiage, and to distinguish it from the earth magnesia'.[21] This is *not* the metal that we now call magnesium; Bergman really is referring to the metal we call manganese! Before we consider the confusion his choice of name caused, we should look more closely at the magnesian earth from which he was trying to distinguish his new metal.

White Manganese and *Magnesia Alba*

Both the loadstone and pyrolusite are usually black, and Bergman uses the name *magnesia nigra*—'black magnesia'—to clarify what he is referring to, since by this time there was also a white magnesia. As we saw earlier, Pliny referred to a white 'magnet' from the Asian Magnesia which didn't attract iron, but exactly what this was isn't known. The white magnesia—*magnesia alba*—is occasionally referred to in early alchemical texts, but again it isn't clear what is meant. This is highlighted in the *Theatrum Chemical Britannicum*, an anthology of early alchemical verse assembled from various ancient manuscripts written in English and complied by the seventeenth-century collector and founder of the Ashmolean Museum, Elias Ashmole. Here he reproduces a poem entitled 'The Hunting of the Greene Lyon', supposedly by a 'Vicar of Maldon', which includes the lines:

To create *Magnesia* they made no care,
In their Bookes largely to declare;
But how to order it after hys creacion,
They left poore men without consolacion[22]

Perhaps the white magnesia was simply a white substance from one of the regions called Magnesia; or maybe it was some white substance that shared certain chemical properties with the other minerals called magnets. For example, whereas the normal form of the mineral used to cleanse glass, pyrolusite, is black or possibly brown (as in Scheele's *brunsten*), it is possible to prepare a white substance from it that will have the same effect on glass. Dissolving pyrolusite in hot, concentrated acid and then adding sodium or potassium carbonate solution yields a white precipitate of what we would now call manganese carbonate. If this substance is heated in air, some pyrolusite is reformed, which is why it could still be used to whiten glass. Scheele prepares manganese carbonate in this way and calls it *weißen Braunsteins* ('white brown-stone'); in the English translation, it is termed 'white manganese'. Manganese carbonate also occurs naturally as the mineral rhodochrosite (often with a pinkish colour), and as a mixed calcium-manganese carbonate in the mineral kutnohorite (named after a region where it is found in the Czech Republic). It is therefore possible that Pliny's 'white magnet' may be referring to a white compound of manganese that could have the usual effects in the manufacturing of glass. Cronstedt actually mentions a white mineral of manganese, *'magnesia alba stricte sic dicta'* ('strictly so-called white magnesia') but adds that it is very scarce.[23]

While the term 'white manganese' has been used for manganese carbonate, it was also used earlier for a completely different compound, what we would now call *magnesium* carbonate. Magnesium and manganese are two completely different elements from the periodic table which have ended up with similar names due to the confusion between their compounds.

One of the first people to study extensively the preparation and properties of magnesium carbonate was the German physician Friedrich Hoffmann, who, in the early decades of the eighteenth century, referred to the substance using the Latin *'magnesia alba'* ('the white magnesia', or 'the white magnet'). The first account of this in English seems to be in a translation of Hoffmann's work entitled *New Experiments and Observations upon Mineral Waters, Directing Their Farther Use for the Preservation of Health, and the Cure of Diseases*, published in 1731. The translator uses the term 'white manganese' for the substance that may be prepared from the liquors left over during the artificial preparation of saltpetre from plant ashes. 'The white Manganese is that chalky, alkaline Matter, obtained by evaporating, and calcining the Remains of the Mother-Liquor, left upon the refining of Salt-petre, that will not shoot into Salt.'[24] Twelve years later, in the second edition of this work, the term *'magnesia alba'* is also included and he adds that 'this white Manganese is but little known, and very little used in England; tho' an agreeable and gently purgative Medicine'.[25] Prior to the French reform of nomenclature when the term 'carbonat of magnesia' was proposed for the substance, it was this Latin, *'magnesia alba'*, that was widely adopted in pharmacopeias and other texts throughout the eighteenth century.

Although initially little used in England, its inclusion and promotion in the influential book *An Essay upon Nursing, and the Management of Children from Their Birth to Three Years of Age* in 1748 meant it was soon to be found in all apothecary shops. We also encountered this *magnesia alba* in Chapter 4 as one of the substances the Scottish chemist Joseph Black investigated in 1756 when studying the absorption and release of carbon dioxide, 'fixed air', from certain carbonate minerals. Black notes the historical production of *magnesia alba* developed by Hoffmann and its by then common use in medicines, but cautiously warns, 'Although *magnesia* appears from this history to be a very innocent medicine; yet…some hypochondriacs who used it frequently, were subject to flatulencies and spasms.'[26]

In this famous treatise on *magnesia alba*, Black compares the properties of magnesium carbonate with the chemically similar calcium carbonate (marble or chalk). One of the characteristic differences between the two is the nature of the solutions left over after dissolving the carbonates in spirit of vitriol (an old term for sulfuric acid). The acid reacts with both carbonates, liberating carbon dioxide and leaving behind a solution of the sulfates. The difference is that calcium sulfate (which is only sparingly soluble in water) has a rather insipid taste, whereas the much more soluble magnesium sulfate has a distinctly bitter taste. Scheele noted that a solution of *manganese* sulfate, formed from sulfuric acid and manganese carbonate, his 'white manganese', also has a bitter taste; it is possible that this common property added to the confusion between the two different substances.

However, it was this bitter salt, magnesium sulfate, that was used by Hoffmann as an easier route to prepare *magnesia alba* without utilizing the solutions required for the manufacture of saltpetre. It was also the bitter taste that led to the discovery of magnesium sulfate about a century earlier, in a small market town located just south-west of London called Epsom.

Magnesium Sulfate—Epsom Salts

According to the local histories of Epsom, now known for horseracing, as well as its mineral salts, the precious spring was discovered in 1618 by one Henry Wicker, who, during a dry summer, observed a small hole in the ground full of water, which he then enlarged so his cattle could drink. However, the cattle would not touch the water, and it was supposed that it contained the well-known salt alum (potassium aluminium sulfate). For a while, the waters were only used to treat external cuts and bruises, but around 1630, some labourers accidentally drank the water and inadvertently discovered its purgative properties.[27] The chemical properties of the water were studied in the 1690s by Fellow of the Royal Society Nehemiah Grew, who published his findings first in Latin, then in English as *A Treatise of the Nature and Use of the Bitter Purging Salt Contain'd in*

Epsom and Such Other Waters. In this small pamphlet, Grew notes that the Epsom salts (magnesium sulfate) reacts with salt of tartar (potassium carbonate) or other carbonates such as the 'urinous salt' (ammonium carbonate) to form a white precipitate (magnesium carbonate): 'A dissolution of this Salt, and Salt of Tartar, and any other Urinous or Lixivial Salt, will generate a white Coagulum, or a Neutral Salt, of the taste of neither, but something Styptick.'[28] This was the preparation later developed and published by Hoffmann that replaced the alternative extraction of *magnesia alba* from the left-over liquors from saltpetre production.

The question remains as to why this substance, magnesium carbonate, came to be called *magnesia alba*. Perhaps, as discussed earlier, it was because it was mistaken for the white manganese, manganese carbonate. The mineral dolomite, for example (a double carbonate of calcium and magnesium), is easily confused with kutnohorite (a double carbonate of calcium and manganese). However, there are other suggestions as to how it got its name, usually to do with the power of attraction, akin to that of the genuine magnet. In the *Art of Glass* from 1699, the author writes, 'The Ancient Philosophers, call also every thing Magnesia, that has a Magnetical Power of Attracting the Occult Virtues of the Heavens and Astral Influences to it.'[29] In his booklet on the origin and nature of *magnesia alba* and Epsom waters from 1767, author Dale Ingram writes that the term '*magnesia alba*' has long been used by alchemists and chemists to describe any white, earthy substances, 'but in particular to express such substances as have a peculiar power of attracting or absorbing a nitrous acid from the air when exposed openly, by which its weight is increased and whiteness heightened'.[30] This is a little odd, since the substance contains no nitrates at all, but perhaps came about since it could be isolated during the production of saltpetre, potassium nitrate, which itself seemed to grow out of the air on the walls of cellars and stables.

In the *Laboratorium Chymicum* by Johann Kunckel (published in 1716, thirteen years after the author's death), Kunckel refers to a *Spiritus Mundi*—a precious liquid which a couple of enterprising entrepreneurs were able 'to collect and to use by means of a certain magnet'. They essentially prepared calcium nitrate by dissolving chalk (calcium carbonate) in nitric acid. They then evaporated the resultant solution to leave the solid nitrate, which was then found to attract water from the air. 'This they abstracted, and they said that the water was a *Spiritus Mundi*. A *loth* of it was valued at 12 *groschen* and was used by high and low.'[31] We would now say that the calcium nitrate is hygroscopic—a substance that readily absorbs water from the air. Regarding the efficacy of the *Spiritus*

Mundi, Kunckel correctly points out 'that the belief must have come to fill the place of the effect—for mere rain water would have worked as well'. The chemist who prepared this calcium nitrate was Christian Adolph Balduin, who we met in Chapter 3, and it was after accidentally heating it too strongly that he discovered his light-magnet, Balduin's phosphorus. While the substance being used as the water-magnet described by Kunckel was derived from calcium carbonate, the same would have worked had magnesium carbonate been used, so perhaps it was a similar idea that led to this substance being named the white magnet, *magnesia alba*.

In yet another booklet devoted to the substance, this time published in Spain in 1750, the author writes that it received its name magnesia 'as the magnet attracts iron, so this attracts harmful humours of the human body, and casts them out entirely: and alba, by its whiteness, because, being well prepared, it is as white as snow'.[32]

However it got its name, it was the term '*magnesia alba*' that gave rise to the name that featured in so many different salts in the system of the mineralogist Bergman, and also in the epochal reform of the French chemists in the 1780s. Bergman used the names *magnesia vitriolata*, *magnesia nitrata*, and *magnesia salita* for the salts we would now call magnesium sulfate, nitrate, and chloride. The French authors used the terms 'sulfate', 'nitrate', and '*muriate de magnésie*', once again clearly expressing the composition of the salts. But Bergman and the French authors disagreed on the corresponding names for the salts of manganese, with the latter using '*manganèse*' but the former confusingly using his '*magnesium*', the term he used for metallic manganese. Alongside their new French names, the Gallic authors also included Latinized names, since they thought 'it would have been incomplete if we had not offered to the learned of all nations, a method to express themselves in an uniform and intelligible manner'.[33] Amazingly, perhaps through deference to Bergman, their recommended Latin names for magnesium and manganese differed only slightly in the terminal vowel sounds: *magnesiæ* and *magnesii*. For example, magnesium chloride was *Murias magnesiæ*, and manganese chloride *Murias magnesii*. Richard Chenevix, in his *Remarks Upon Chemical Nomenclature, According to the Principles of the French Neologists*, published in 1802, did not hesitate to point out how inappropriate this was, writing: 'The errors of the French Nomenclature are not confined in their own Language.'[34]

As if matters could not get any worse, in 1782 the Irish mineralogist Richard Kirwan introduced the term 'muriatic earth' as an alternative name for magnesia—the adjective 'muriatic' being derived from Latin and indicating that the salts could be obtained from sea water. The problem was that the adjective was also in use for an acid—muriatic acid, which we now call hydrochloric acid. Thankfully, Kirwan's suggestion was only picked up by a few English authors, leading one text to mention muriatic salts (chlorides) and muriatic earth (for compounds of magnesium) on the same page. Fortunately, there seems to be no mention of 'muriate of muriatic earth', which, logically, would have been the name for magnesium chloride.

Kirwan's fellow countryman and occasional collaborator Stephen Dickson, whose caustic criticisms of the new nomenclature we encountered earlier, wrote: 'Neither has the name muriatic earth any title to supersede magnesia.' He correctly pointed out that 'although it be true that this substance is, in certain specimens, a marine or muriatic earth, being found in the sea, yet it has neither an exclusive nor a pre-eminent claim to that epithet. It is not the only earth, or even the most abundantly contained in the sea; for muriated lime is always discoverable in sea-water, and in salt springs; and there is more lime than magnesia mixed with sea salt...'[35] He adds that magnesia is not even found principally in the sea, but mainly found in the water from wells, such as that at Epsom. He concludes, 'Moreover, if this name should be selected for the purpose of expressing a compound of magnesia and some other substances, by using the adjective muriated, this contrivance would be utterly unjustifiable; for the adjective belongs by prior, and therefore exclusive, right to muriatic acid.'

The Heavy Earth—Barytes

Continuing from Cronstedt's early analyses, Scheele recognized that new elements were to be found in some of the minerals Cronstedt studied. In his classic paper published in 1774 on the mineral brunsten, Scheele described certain crystalline impurities and carefully detailed their chemical properties. In addition to a little lime (calcium carbonate) present in the brunsten, he also noted a new earth with subtly different properties. At the time he simply referred to it as a 'peculiar kind of earth', but Scheele's contemporary John Gahn (the man we met earlier as being the first to isolate metallic manganese) had also noticed this earth in the mineral Cronstedt had called *tungspat*, the Swedish for 'heavy spar'. The term 'spar' has long been in use for a mineral that may easily be broken or cleaved into crystalline regular shapes; as Cronstedt wrote, 'when a stone breaks into a rhomboidal, cubical, or a plated form, with smooth and polished sides, it is called spar'.[36] Heavy spar, now known to be barium sulfate, was for some time prior to the work of Scheele, and Gahn thought to be a modification of lime and had also been known as *marmor metallicum* (metallic marble) since it had some resemblance to marble (another form of calcium carbonate) but was much denser, like a metal. When Bergman reported Gahn's work in 1775, he called the new substance the 'earth of heavy spar' and later, '*terra ponderosa*', 'ponderous earth'.

In his reform of chemical nomenclature in 1782, Guyton de Morveau stated 'that every substance should be denoted by a name, and not by a phrase'[37] and so was not satisfied with the 'improper and prolix expressions' 'heavy earth', or 'the terrestrial base of heavy spar'. He proposed instead a word based on the Greek for heavy, at first suggesting 'barote' and its adjective 'barotic'. This was later modified by Kirwan to 'barytes', which carried the approval of the French authors in their nomenclature of 1787, where Guyton wrote, 'we adopt the word *barytes* from βαρύς [barys], *gravis*, which retains enough of the former denomination [heavy earth] to assist the memory, and which differs sufficiently from it so as not to cause a false idea'.[38]

Heavy Stone—Wolf-foam or Jupiter's Wolf

'Heavy earth' or *tungspat* was for some time confused with another mineral mentioned by Cronstedt—'heavy stone', or in his native Swedish, '*tungste*'. The mineral tungsten was another of those analysed by Scheele which he recognized as containing a new element. In light of this, the mineral, now known to be composed of calcium tungstate, was eventually renamed scheelite. But there was another tungsten-containing ore which was known from a much earlier date—*wolfram*. This German word seems to be first mentioned by the mining-preaching minister we encountered in Chapter 2, Mathesius. He writes in 1562: 'Wolfrumb, which the Latins call Wolffschaum [Wolf-foam or wolf-froth], several others call Wolffshar [wolf's hair], as it is black and lengthy, is to be found next to tin ore as galena is to be found next to silver ore.'[39]

The reference to the Wolf-foam is clarified by Petrus Albinus, the sixteenth-century professor at Wittenberg University (where the theologian Martin Luther had taught a few decades before). Albinus describes a substance found with tin ore that 'robs the tin in the fire, making it brittle and patchy.' He adds: 'the Latins call it *spumam lupi* from the little German word Wolffram or Wolffschaum though some think it is called the same as Wolffromm'.[40] This ties in with a much earlier reference made by the father of mineralogy, Georg Agricola, who writes in 1546: 'A certain black stone is found with a uniform colour similar to the stone from which tin is smelted but so light that one readily perceives that it is barren and contains no metal. We call this *spuma lupi*.'[41]

Despite the same (albeit Latinized) name of the mineral 'wolf-foam' and its association with tin, it seems likely that Agricola is referring to a different substance from wolfram since he clearly states his substance is not dense, whereas the tungsten ore is strikingly so. Our other sixteenth-century authority, Ercker, also mentions the ore, which he states 'the old Miners have not known',[42] and refers to it in the German text as 'Wolffram', 'Woffram', or 'Wollferam'; in the English translation of 1683, the variations 'wolfram', 'woolfrain', 'woolferan', and 'wolferan' appear.

The miners relied on being able to refine tin ore by utilizing its greater density compared with that of the accompanying rock matrix, and this is why the wolfram posed a problem. It was not possible to separate the wolfram from the tin ore in the slurry tanks, and the wolfram had a disastrous effect on the quality of the tin produced, making it more brittle and decreasing its value—as Albinus said, 'it robs the tin'. The German mineralogist Johann Friedrich Henckel, a former student of the phlogistian Stahl, wrote in his *Lessons of Mineralogie or Science*, published in German in 1747, that 'wolfram (*lupus Jovis*) is a bad sort of mineral: it does not, as mineworkers imagine, consume tin, but spoils it; it makes it hard because of the iron that it contains'.[43] Henckel also seems to be the first who calls this ore *Lupus Jovis* ('Jupiter's Wolf'), referring to the ancient association between tin and Jupiter.

In recent texts, it is often stated that wolfram received its name from the fact that early miners thought it 'ate up the tin as the wolf eats the sheep'. Sadly, this romantic description seems to originate from the late nineteenth century although in the *New and Curious Mining Lexicon* published in German in 1730, after speaking of the difficulty of its separation and how it spoils the tin, the author does say that perhaps wolfram received its name 'as it steals and devours like a wolf'.[44]

The wolf association is maintained to this day since it is the alternative name *wolfram* that gives rise to the symbol W, used for the element tungsten. The first people to isolate the metal are usually identified as the Spanish brothers Juan José and Fausto d'Elhuyar. The older brother Juan José had visited Bergman and Scheele just after the latter had published his findings on the mineral tungsten. Scheele may also have prepared a crude form of the metal, but the d'Elhuyar brothers published a detailed account which appeared, translated into English, in 1785 as *A Chemical Analysis of Wolfram and Examination of a New Metal, Which Enters into Its Composition*. There is a bit of a mix-up regarding the name the brothers chose for their new metal. In the original French version, they stated that they chose the name *Volfran*, using the ending '-*n*' to distinguish it from the mineral *Volfram*. They preferred this to the name *Tungste* or *Tungstene* derived from the metal's other common mineral, since wolfram had been known for much longer. However, despite noting the change of '-*m*' to '-*n*', in the English version, the translator gives the new name as wolfram. It is probably because the French authors chose *tungstène* for the metal in their nomenclature of 1787 that this also came into English, initially as 'tungstein'. Berzelius based his chemical symbols on the Latinized names of the elements, and had decided in 1811 that 'wolfranicum is without doubt a bad name, but it is the best among those that have been given to this metal; because the one we have derived from the Swedish word, tungsten, which means a heavy stone, is even less proper'.[45] As we saw in the previous chapter with the names for sodium and potassium, the Scottish chemist Thomas Thompson tended to ignore the great Berzelius, making his own changes when he published the Swede's paper on chemical symbols. Although at one point Thompson did use the symbol W, based on wolfranium, later in the same paper he changed it to Tn for tungsten. Thankfully a universal standard was eventually agreed, and chemists worldwide now use W for the element tungsten.

Leady Confusion

In addition to his work on the earths of manganese, barium, and tungsten, it was Carl Wilhelm Scheele once again, who, by identifying a new element, resolved another area of confusion between three completely different minerals that had been thoroughly mixed up for hundreds of years. These are the minerals galena (lead sulfide), graphite (a form of the element carbon), and molybdenite (molybdenum sulfide). The confusion was so severe that we still use terms derived from the element lead for names used for all three of these substances.

Just as with loadstone and pyrolusite, all three of these minerals were mixed up because of their remarkably similar appearance—in this case, a dark bluish-grey shiny substance. By far the most common of the three is what we would now call galena, lead sulfide, so it is not surprising that the less common minerals were confused with this and ended up with names related to lead species. The Greek word for 'lead' is '*molybdos* (μόλυβδος)', and the word '*molybdæna*' was used for things associated with lead, such as the plummet weight used in the builder's plumb-line. The same word '*molybdæna*' was also used for the lead oxide scum formed during the purification of silver ores. This scum was also known as *spuma argenti* (meaning 'foam of silver'), *litharge* (derived from Greek words for 'stone' and 'silver'), and *plumbago* (perhaps signifying a sort of lead rust). In his section on lead, Pliny uses all four of the terms '*molybdæna*', '*galena*', '*spuma argenti*', and '*plumbago*'.

Although Pliny seems to use these words mainly for different variations of lead oxides—which are usually found as brightly coloured reds and yellows, and may have been some of the first compounds of lead mined—by the time of Agricola in the sixteenth century, the most common source of lead was the lustrous black mineral we now call galena (lead sulfide). Agricola suggests that Pliny actually means this when he used the terms '*galena*', '*molybdæna*', and '*plumbago*'. The fact they are so strikingly different is picked up in Agricola's introductory text that we first encountered in Chapter 2, *Bermannus*, written in 1530 as a dialogue between the mineralogist 'Bermannus' and the scholars 'Nicolaus Ancon' and 'Johannes Neavius'.

> Naevius: [...] there is still one thing that bothers me.
>
> Bermannus: What is it? I shall see if it is possible to explain it.
>
> Naevius: Dioscorides writes that his mineral molibdaena found at Sebastia near Corycos is golden-yellow and brilliant yet the mineral you show me has a certain brilliancy but is lead-grey and by no means golden-yellow.[46]

The Master replies that the student is 'at liberty to agree or disagree', but he thinks that Dioscorides uses a different term, essentially 'lead-stone', to refer to the lead-coloured galena. Agricola also mentions a 'sterile' variety of galena—one that is completely consumed by the fire, yielding no lead. This is most likely a reference to graphite, which when heated strongly in air forms gaseous carbon dioxide and leaves only a residue of its trace impurities.

Graphite—Black Lead

The first unambiguous descriptions of graphite are from the late sixteenth century, when its use by artists and writers is noted. In one of his sermons, Johann Mathesius includes a summary of the writing implements used up to his time, finishing with a new 'metal' that must surely be graphite:

I think that one still writes on wax tablets with iron styluses, which was very common of old, afterwards one used to write with silver styluses on white, wooden boards or tablets, or with lead on varnished parchment and with ink on vellum, and now on slate tablets with a slate pencil, or on paper with a new and self-growing metal.[47]

Particularly famous was the English graphite from the Borrowdale mine near Keswick in Cumberland, and it was probably a sample from here that was referred to in the first description and illustration of a pencil in 1565. The mines were included in the 1610 English edition of William Camden's *Britain*, or, to give it its full title, *Britain, or A Chorographicall Description of the Most Flourishing Kingdomes, England, Scotland, and Ireland, and the Ilands Adioyning, out of the Depth of Antiquitie*. Here the author describes where 'the river *Derwent* hideth himselfe in the Ocean; which having his first beginning in *Borrodale*, a valley hemmed in with crooked hilles, creepeth betweene the mountains called *Derwent Fels*'. In addition to copper mines, 'here also is commonly found that mineral kind of earth or hardned glittering stone (we cal it Black-lead) with which painters use to draw their lins & make pictures of one colour in their first draughts'. Not surprisingly, Camden is not sure exactly what the mineral is, and is content to 'let others for me search it out.'[48] The term 'black lead' had earlier been used for the metallic lead itself, to distinguish it from the metals tin and bismuth (see Chapter 2). Here, though, it is used for a mineral which is distinctly lead-free but, like its namesake, could easily be used to mark paper—or indeed sheep, as the locals from Borrowdale were wont to do.

The Borrowdale mine is also mentioned in Webster's *Metallographia: or, an History of Metals* from 1671:

> Here it cannot be amiss to say something of that which we commonly call Black-Lead, because it discoloureth the hands far more then common Lead, and is that whereof Pencils are made for Painters and Scriveners, and many other such like uses. In the North we usually call it Kellow, and some call it Wadt; of which there is still a Mine near Keswick in Cumberland, which is opened but once in eight or ten years; either by reason of its scarceness, or to keep up the prices of it…[49]

The 'kellow' mentioned, with its alternative spelling 'killow', seems to be a local word also used for similar soft, black, graphitic minerals and is perhaps related to 'collow', a sooty grime of coal or coal-dust. 'Wadt' may also be found with the variant spellings 'wad' and 'wadd', and probably meant 'black'. Interestingly, later in the eighteenth century this word became more commonly associated with the black manganese oxide, the brunsten famously studied by Scheele. It is possible that the phrase 'wadt-lead' may have given rise to an early German term used for graphite—'Wasser-blei' ('water-lead')—but it is also possible that the reverse could be true. The term 'wasser-blei' might also be related to another name given to graphite, especially in France: le Plomb de Mer ('sea-lead'), or, in Latin, Plumbum marinum. However, when talking of black lead in the section on 'Lead Oars', the seventeenth-century French pharmacist Pomet says, 'The Ancients gave it the Name of Plumbago, and of Sea Lead, because they pretend they took it from the Bottom of the Sea.'[50] He adds that black lead of the finest quality is used to make 'the long Pencils that are so much exteem'd' and consequently 'Lead of these Qualities wants for no Price, the Marchant may have what he pleases, being much sought for by Architects and other Persons for drawing.' Even today, we call the graphite core of a pencil the 'lead', despite none of this metal being present.

In addition to all the lead-based terms used for the mineral graphite, the confusion was enhanced with the inclusion of the word 'molybdæna'. This is the term used in the 1741 English edition of Cramer's Elements of the Art of Assaying Metals, and when he discusses the mineral molybdæna, Cramer notes that it is 'otherwise called Cerussa nigra, Plumbum marinum, in English Wad or black-lead, in German Wasser-Bley'. 'Cerussa nigra' here means 'black-ceruse', ceruse being a white carbonate mineral of lead. Despite all these names derived from lead, Cramer does state that the molybdæna 'must not be confounded with the Galæna, or Steel-grained lead-Ore, which though commonly called by the same Name, yet is altogether different from it'.[51]

Galena is the name for the mineral compound in which lead is chemically combined with sulfur, and both of these elements may easily be extracted from it. While neither lead nor sulfur may be extracted from pure graphite (which is simply a form of the element carbon), further misunderstandings arose since the third species in our triad of confusion, now known as molybdenite, has the same appearance as both graphite and galena, and is like graphite in that no lead can be extracted from it, but like galena in that it contains sulfur. One of the first to study a sample of what was clearly molybdenite (although he still thought it was graphite) was yet another Swedish mineralogist, a master smelter and assessor of mines, Bengt Andersson Qvist. Although Qvist correctly found his sample contained much sulfur, he mistakenly thought it was a compound of iron and tin. His work, published in 1754, was picked up four years later by Cronstedt in his seminal book on mineralogy. Under a section on 'Sulphur that has dissolved, or is saturated with metals', Cronstedt includes this supposed compound of iron, tin, and sulfur, which he gives in the Swedish original the names *Blyerz* (lead-ore) and *Wasserbley*, and which in the English translation are given the names Black Lead and Wadd, and significantly, in both editions, the name Molybdæna.

Molybdenum and Graphite Revealed

It was Scheele who was to bring order to the chaos by revealing the true nature of graphite and by recognizing the new element in molybdæna. He published his findings in two papers: *Experiments with Lead-Ore: Molybdæna* (translated from the Swedish original) came in 1778, with *Experiments with Lead-Ore: Plumbago* following the year after. The titles of these two papers are significant; neither is actually concerned with lead ore, but both help to settle the meanings for the other term mentioned. The English translation that appeared in 1786 makes no mention of lead-ores, and Scheele starts off by saying, 'I do not intend to treat here of the common molybdæna [lead-ore] which is to be met with in the shops, for that is very different from the sort concerning which I am now communicating my experiments to the Royal Society.' He continues: 'Mine is that kind which in Cronstedt in his mineralogy is called *Molybdæna, membranacea, nitens*, and with which Quist and several others made their experiments.'[52]

We would now call the mineral Scheele was experimenting with molybdenite, composed of molybdenum disulfide, MoS_2. On treating this mineral with nitric acid, Scheele obtained a chalk-white powder which he called *terra molybdæna*, or earth of molybdæna, and which we would now call molybdenum trioxide, MoO_3. After accurately describing various reactions of molybdenite, Scheele wanted to confirm its constitution by resynthesizing the graphite lookalike. He writes: 'Having now analysed molybdæna, by means of the experiments which I have communicated, it still remains to be able to recompose this mineral of its proximate parts. That molybdæna contains sulphur, is already known, and my experiments shew the same thing.' He then describes heating sulfur and his earth of molybdæna to obtain 'a black powder, which, when rubbed between the fingers, stained them of a shining black colour, and shewed the very same phænomena in every other respect, as native molybdæna itself'. He correctly concludes that 'we have then a kind of earth in molybdæna, which has probably to this time been unknown, and which one may properly call acid of molybdæna, as it has all the properties of an acid'.[53]

Despite his attempts, Scheele was not able to obtain a sample of the pure metal from the ore (although actually, he probably did, but it was just very finely divided and appeared as a black powder rather than as a metallic lump). Realizing that he probably needed a stronger furnace to produce the metal, he asked his friend the mineralogist Peter Jacob Hjelm to try. Hjelm succeeded in 1781, and following Scheele's example, gave his 'perfect molybdænic metal' the name 'molybdænum'. Even though the name of the metal was derived from the Greek word for the metal lead, it was this original version 'molybdænum', or simply 'molybdenum', that stuck—despite the occasional flirtation with 'molybdenium'. Even Berzelius accepted 'molybdænum' and gave the element the symbol that we still use today, Mo.

Scheele's paper on graphite was a showcase of his virtuosity as an analyst—and perhaps of his inability or indifference in choosing appropriate names for his discoveries. In Swedish, his paper was entitled *Försök med Blyerts, Plumbago* best translated as *Experiments with Lead-Ore: Plumbago*; but the English translation of 1786 was simply *Experiments on Plumbago*. Since the paper is about graphite, a form of carbon, it is curious that Scheele starts by referring back to his first comments on his earlier paper, 'as I stated there at the outset that the lead-ore generally known in commerce was very different from molybdaena of which I there treated, I now have the honour to prove this by experiments'.[54] Throughout the paper in the original Swedish, he refers to graphite as 'Blyerts'—'lead-ore'—despite going on to show that the mineral does not contain any lead! In the English translation of the day, the translator does not use the term 'lead-ore' and instead writes that 'the black lead or plumbago which is generally known in commerce, is very different from molybdaena'.[55] The translator then continues to use the word 'plumbago' throughout, thereby cementing that word in the dictionary to mean (usually) 'graphite'.

As usual, Scheele is spot on with his analysis. On heating with what we would now call oxidizing agents—on most occasions he used nitre (potassium nitrate)—Scheele found carbon dioxide gas was produced. He collected the gas, which he called aerial acid, in a large ox-bladder in order to study it further. Scheele gives his conclusion in terms of the then-common phlogiston theory, writing: 'Hence I consider myself satisfied that plumbago is a kind of mineral sulphur or charcoal, of which the constituents are aerial acid united with a large quantity of phlogiston.'[56] In modern terms, this is equivalent to saying that graphite consists of carbon dioxide minus the oxygen. Scheele also correctly realizes that the graphite usually contains a portion of iron pyrites (iron disulfide) as an impurity. This caused problems for later investigators who thought that the iron was an integral part of the graphite. The idea that graphite was a 'carburet of iron' persisted for around thirty years, until it was finally realized that Scheele's account was quite correct and that the iron pyrites was only an impurity.

After Scheele had accurately demonstrated exactly what the substance was, it was recognized that a name based on the metal lead was far from ideal. Consequently, in 1789, Abraham Gottlob Werner, who has been called 'the father of German geology', gave it the new name graphite from the Greek 'graphein' ('γράφειν'), 'to write', because of its most common use in pencils.

Scheele's thorough investigations of these new 'earths' eventually led to the isolation of the metals manganese, tungsten, molybdenum, and barium. Scheele saw the extraction of the first three of these, and they are included among the seventeen metals listed in Lavoisier's *Table of Simple Substances* that we met in Chapter 4 (Figure 33). Neither Scheele nor Lavoisier lived to see the isolation of barium—but the earths Barytes, Magnesia, Alumina, and Silice were also included in the *Table*, even though Lavoisier and his collaborators 'presume[d] that the earths must soon cease to be considered as simple bodies'.[57] It was to be Davy who, not twenty years later, isolated metals from the alkalis and earths; but before he did this, Lavoisier's prophecy inspired a premature false claim to the extraction of metals from the earths, and with this claim came new names which were thought to be quite inappropriate.

Austrum, Borbonium, and Parthenum

In the second English translation of Lavoisier's *Traité*, the translator Robert Kerr updated the section on earths with the news that, 'In the laboratory of the Academy of the mines at Chemnitz in Lower Hungary, some experiments have been lately made, by Messrs Tondi and Ruprecht, by which the number of the metals seems to be considerably augmented.'[58] These two scientists—Italian Matteo Tondi, who initially studied at the University of Naples, and Anton Leopold Ruprecht, who flourished in eighteenth-century Austria/Hungary—in addition to 'ascertaining the real metallic nature of Tungstein, Molybdena and Manganese, which some chemists had doubted...have succeeded in procuring metallic reguli from Chalk, Magnesia, and Barytes.' The findings were initially published by one of Tondi's students, who writes that Lavoisier was the inspiration behind the research: 'Encouraged by his successful reduction of tungsten and manganese, and further stimulated by Mr Lavoisier's theory and reasoning relative to the reduction of metals in general; Mr Tondi resolved to try whether, by means of the same process, the simple earths could be reduced, and thus to see whether the conjecture which Mr Lavoisier had thrown out in his Elementary Treatise, that all the earths are probably metallic substances, was just or not.'[59]

The experiments all involved heating the earths with a mix of linseed oil and charcoal, and gave metallic globules which in some cases were attracted by the magnet. The great German analyst Klaproth soon proved that these were largely samples of impure iron formed from the clay crucibles the reactions were carried out in. However, the continental 'discoverers' proposed new names for their metals, as Kerr describes: 'To these three new metals Mr Tondi wishes to give the names of *borbonium*, for the regulus of barytes; *austrum*, for the regulus from magnesia; and *parthenum*, for that of chalk.'[60] These names reflected the patriotism of their discoverers, being derived from Parthenope (the site of the ancient Greek colony that later became the 'New Town' Neapolis or Naples); the House of Bourbon—the royal dynasty who ruled over Naples at the time; and Austria. To Kerr, these new names were not appropriate:

> It were hard to deny a discoverer the right of giving names to his own discoveries, without some reasonable objection; but these names would introduce confusion into chemical nomenclature, which it has been the great object of the French chemists to reform, and render regular; wherefore I would propose that they should be named *barytum*, *magnesium*, and *calcum*: These accord with the reformed old names of the substances from which they are procured, merely by changing to the neuter gender, in which all the names of the metals are placed in the new nomenclature, and then the three, formerly called, earths will be oxyds of these metals respectively, or *baryta*, *magnesia*, and *calca*, if single terms are preferred, these latter being in the feminine gender, which is appropriated to alkaline substances in the new nomenclature.[61]

It is curious that even though Kerr mentions Klaproth's refutation of these discoveries, he does not take this section out of the subsequent English editions of Lavoisier's work; in fact, in the fifth edition of 1802, Kerr even includes the symbols Ca, Mg, and Ba for the 'newly discovered metals' in his translation of the symbols of Hassenfratz and Adet, which he includes for the first time in the appendix (Chapter 5, Figure 41).

Davy's Elements

Calcium, magnesium, barium, and strontium were finally isolated by Humphry Davy in 1808, the year after he first prepared metallic potassium and sodium.

The earths were trickier to isolate, as he could not use exactly the same method he used before—electrolysing the molten hydroxides of potassium or sodium. The problem with the earths was that heating their hydroxides simply drives out the water to form the oxides (the earths proper), which then could not be melted and so would not conduct the electrical current necessary to split them into their component elements. Davy's early attempts involved electrolysing mixes of the earths with mercury oxide, and this afforded tiny quantities of amalgams—mixtures of the metals with mercury. Davy then perfected his method after hearing how Berzelius had obtained better results using mercury as one of the electrodes. Davy repeated the experiments by passing the current from the giant battery at the Royal Institution through a slightly moistened mixture of mercury oxide and the appropriate earth, piled up on a plate made of platinum (which served as the positive electrode), with the negative electrode being inserted into a little globule of mercury recessed in the top of the mix. Unlike Berzelius, Davy managed to get sufficient quantities of the amalgams to be able to distil off the mercury and leave the new metals behind—although often still contaminated with mercury, especially in the case of calcium.

Davy had to give names to the new metals, but since Bergman had already used the name magnesium for the metallic form of manganese, Davy felt he had to come up with a different one. He writes: 'These new substances will demand names; and on the same principles as I have named the bases of the fixed alkalies, potassium and sodium, I shall venture to denominate the metals from the alkaline earths barium, strontium, calcium, and magnium; the last of these words is undoubtedly objectionable, but magnesium has been already applied to metallic manganese, and would consequently have been an equivocal term.'[62] However, by the time he published his book *Elements of Chemical Philosophy* in 1812, he had been persuaded to change his mind, writing in a footnote: 'In my first paper on the decomposition of the earths, published in 1808, I called the metal from magnesia, magnium, fearing lest, if called magnesium, it should be confounded with the name formerly applied to manganese. The candid criticisms of some philosophical friends have induced me to apply the termination in the usual manner.'[63]

Fiery Plutonium

Perhaps because of the confusion between magnesium and manganese, the former, at least for a while, was referred to as *talcium* or *talkium* in Germany. This derives from another common magnesium-containing mineral, talc—a form of hydrated magnesium silicate. However, a suggestion that Davy vehemently objected to was an alternative proposal to his name 'barium'. A professor of mineralogy at the University of Cambridge, Edward Daniel Clarke, thought he had isolated samples of the alkaline earth metals using the extremely powerful oxy-hydrogen blowtorch. He seems to have been rather fond of the classics, since he began his paper, 'If the chymists of former ages had been told that to increase the action of *fire* it is necessary that the combustible be *water*, some such author as *Agricola*...would perhaps have maintained that this truth was mystically typified in the rape of *Proserpine*, by *Pluto*, from the fountain of *Cyane*.'[64] A favourite subject of artists of the Renaissance, the story concerns the abduction of the goddess Persephone, daughter of Zeus and Demeter, by Hades or Pluto, the god of the underworld. As we have seen, the name barium was derived from the heavy earth investigated by Scheele. Clarke objected to this name on the basis that while the mineral itself is dense, the free metal is not particularly so. He writes: 'As it will be necessary to bestow some name upon it, and as any derivative from βαρυς [barys] would involve an error, if applied to a *metal* whose *specific gravity* is inferior to that of *Manganese* or *Molybdenum*, I have ventured to propose for it the appellation of PLUTONIUM; because we owe it entirely to the *dominion of fire*. According to *Cicero* there was a *temple* of this name, dedicated to the *God of Fire*, in *Lydia*.'[65] Davy was not happy with this proposed change of name, and apparently considered it a personal attack. He wrote to Clarke: 'I cannot agree with you as to the propriety of altering the name of the metal of barytes...Your expts furnish no *new* reason for altering the general principles of Nomenclature adopted when I first discovered the decomposition of the alkalies & earths.'[66]

Davy's Failures—Ichthyosauros Cutlets

Davy was unable to prepare samples of the other earths known at that time, but suggested names for them nonetheless, writing: 'Had I been so fortunate as to have obtained more certain evidences on this subject, and to have procured the metallic substances I was in search of, I should have proposed for them the names of silicium, alumium, zirconium, and glucium.'[67] The only one of these names to have survived intact is zirconium. The new element was first identified in the form of an earth (that is, the element combined with oxygen) by Klaproth in 1789, the same year he discovered uranium. He found the new earth in a precious stone from Ceylon (now Sri Lanka) called 'jargon of Ceylon', circon, and later, zircon. He writes: 'I think myself justified in considering it as a *new, distinct, simple earth*, before unknown; and at present I give it the name of *Zircon-earth (Terra Circonia)*, until it may, perhaps, be found in other species of stones, or possessed of other properties, that may give rise to a more appropriate denomination.'[68] As it turns out, it was soon found (by himself) in another, more common stone, known as the hyacinth. But this gave him a dilemma—which stone should give its name to the earth? He writes, 'The jargon has, indeed, already obtained that distinction; but ought it not to be transferred to the hyacinth, being a gem much older, longer known, and more esteemed?—If so, the denomination *hyacinth-earth* should then be adopted, and substituted to that of *circonia*, or *jargonia*.'[69] As attractive as the name hyacinthium might have been, it was, of course, zirconium metal that was first isolated by Berzelius in 1824, by heating potassium metal with a salt of Klaproth's earth.

Berzelius is also usually credited with the first isolation of silicon in 1823 by the same method—this time heating potassium metal with silicon fluoride—but others before him had also tried this, with varying success. Berzelius was reluctant to change the name silicium, but Scottish chemist Thomas Thomson had objected to the name, writing, 'The base of silica has been usually considered as a metal, and called *silicium*. But as there is not the smallest evidence for its metallic nature, and as it bears a close resemblance to boron and carbon, it is better to class it along with these bodies, and to give it the name of *silicon*.'[70]

As we saw earlier, Davy's glucium, although for a while known as glucinium, eventually became beryllium, and both beryllium and aluminium metals were also first obtained by heating their anhydrous chlorides with potassium metal. Davy's name alumium deserves further comment. While his name perhaps reflects better the metal's origin from the salt alum (see earlier), it did not exactly roll off the tongue. Davy himself soon changed his proposed name to aluminum, and then finally to aluminium. For a while, both were in common use on both sides of the Atlantic, although 'aluminum' has since become the preferred American term. Initially it hardly mattered, since the element could only be prepared in tiny quantities. In fact, aluminium was initially so rare and expensive—even more so than gold—that Emperor Napoleon III had some cutlery made from the strikingly light metal to use at state banquets. But when the element could finally be produced cheaply in larger quantities, it came to the attention of the general public, and its name was questioned. During the 1850s, Charles Dickens used to oversee the production of a weekly magazine *Household Words*. In the issue dated 13 December 1856, an article entitled 'Aluminium' announced, 'In a short time we shall be in possession of a new metal, which need not be ashamed to announce itself by a distinct name.'[71] After giving a brief history including mentioning Lavoisier's brilliant insight into the probable existence of new metals in the alkalis and earths, the author writes:

> What do you think of a metal as white as silver, as unalterable as gold, as easily melted as copper, as tough as iron; which is malleable, ductile, and with the singular quality of being lighter than glass? Such a metal does exist, and that in considerable quantities on the surface of the globe. 'Where? From what distant region does it come?' There is no occasion to hunt far and wide; it is to be found everywhere, and consequently in the locality which you honour with your residence. More than that, you do not want for it within-doors at home; you touch it (not exactly by direct and simple contact) several times in the day. The poorest of men tramples it under his feet, and is possessed of at least a few samples of it. The metal, in fact, in the form of an oxide, is one of the main component elements of clay; and as clays enter into the composition of arable land, and are the material on which the potter exerts his skill, every farmer is a sort of miner or placer, and every broken potsherd is an ingot in its way. Our new-found metal is ALUMINIUM.[72]

After extolling the wonders and merits of this new metal—'Henceforward, respectable babies will be born with aluminium spoons in their mouths'—the authors come to the choice of name for the metal:

> A final word. If aluminium is hoping to replace either gold and silver, or copper and tin, or to take its own place without replacing anything, it may do so in the Arts and manufactures; but it never can in literature or popular speech, unless it be fitted with a new and better name. Aluminium, or, as some write it, Aluminum, is neither French nor English; but a fossilized part of Latin speech, about as suited to the mouths of the populace as an ichthyosauros cutlet or a dinornis marrow-bone. It must adopt some short and vernacular title. There would be no harm in clay-tin, while we call iron-ware tin; loam-silver might plead quicksilver, as a precedent; glebe-gold would be at least as historically true as mosaic gold. A skilful word-coiner might strike something good out of the Greek and Latin roots—argil, though a Saxon etymology is far preferable. But something in the dictionary line must be attempted. I should like to know what will become of poor 'Aluminium' when it gets into the mouths of travelling tinkers or of Hebrew dealers in marine stores?[73]

It was not until 1990 that the International Union of Pure and Applied Chemistry decreed that the official name should be aluminium, although it was by then too late to make much difference to the American usage of 'aluminum', which had been adopted by the American Chemical Society thirty-five years earlier.

There was one other element that Davy could not isolate, despite numerous attempts to do so, one of which nearly cost him his life: fluorine. Although he did not isolate this most reactive of all the elements, Davy did help in determining its name. To understand how, we first need to look at one of the element's relatives, an element that Davy did eventually name even though he did not discover it. This is the element chlorine.

7

THE SALT MAKERS

'Why, Uncle! do you really mean that green smoke came out of salt—the salt that we eat?'
—Rider Meyer, 1887[1]

This chapter looks at the elements from the penultimate group of the periodic table—the halogens ('salt-formers'). We shall see that the first of these elements was discovered by Scheele during his investigations of the mineral pyrolusite. Lavoisier knew of the element but he failed to recognize it as such since he was convinced the gas had to contain oxygen and so must be a compound. It was left to Davy to prove that this was not so, which led to the English chemist naming this element that had been discovered (but not properly named) over thirty years before by the great Scheele. Davy's choice was to influence the names given to all the members of this group, including the most recent member named in 2016.

Marine Acid Air

There are three common acids known as mineral acids, since they may all be obtained by heating combinations of certain minerals. Their modern names are nitric acid, sulfuric acid, and hydrochloric acid. Of these three, hydrochloric was probably the last to be discovered. Nitric and sulfuric acids were obtained in the thirteenth or early fourteenth centuries, but the earliest unambiguous preparation of relatively pure hydrochloric acid is from a hundred years later, in a manuscript from Bologna which translates as *Secrets for Colour*. It gives a curious recipe for a water to soften bones: 'Take common salt and Roman vitriol in equal quantities, and grind them very well together; then distil them through an alembic, and keep the distilled water in a vessel well closed.'[2] As we saw in Chapter 3, 'Roman vitriol' is a hydrated metal sulfate, probably iron or copper sulfate; its mixture with salt, when heated, produces water and hydrogen chloride, which together form the acid solution. Later texts from the sixteenth and seventeenth centuries include similar methods to prepare this so-called spirit of salt, or 'oyle of salt'. The first mentioned use, to soften bones, is indeed best achieved with hydrochloric acid, which readily dissolves the minerals from bone to leave only the organic matter largely intact. Leave a chicken bone in dilute hydrochloric acid for a few hours, and it may easily be bent without breaking. The residual organic matter, sometimes known as 'ossein', is far more likely to be damaged by the other mineral acids.

While solutions of the acid had been known since the fifteenth century, it was not until the late eighteenth century that pure hydrogen chloride gas was isolated. It was at this point that Joseph Priestley became intrigued by an observation by Cavendish, who mentions an 'elastic fluid, which retains its elasticity as long as there is a barrier of common air between it and the water, but which immediately loses its elasticity, as soon as it come into contact with the water'.[3] Cavendish was studying the reactions of different acids and metals to produce hydrogen (see Chapter 4), but when he tried heating spirit of salt (hydrochloric acid) with copper, no hydrogen was produced; just this extremely water-soluble 'elastic fluid' we now call hydrogen chloride gas. Priestley soon found that the copper actually played no part in this reaction, and that the gas was simply liberated on heating the acid. Furthermore, he realized he could collect the gas in the same way he had collected the other extremely water-soluble gas, ammonia, by collecting it over mercury instead of water. Priestley notes: 'this remarkable kind of air is, in fact, nothing more than the vapour, or fumes of spirit of salt, which appear to be of such a nature, that they are not liable to be condensed by cold, like the vapour of water, and other fluids, and therefore may very properly be called an *acid air*, or more restrictively, the *marine acid air*'.[4] The free gas readily dissolves in water again, to reform the hydrochloric acid. In fact, Priestley found that 'water impregnated with it makes the strongest spirit of salt that I have seen, dissolving iron with the most rapidity'. He adds that in comparison, 'two thirds of the best spirit of salt is nothing more than phlegm or water'.

Priestley used the term 'marine acid', but many others, including the French reformers, preferred the Latinized term 'muriatic acid'. But the problem was that they were not sure what the acid actually was. Lavoisier writes, 'Although we have not yet been able, either to compose or to decompose this acid of sea-salt, we cannot have the smallest doubt that it, like all other acids, is composed by the union of oxygen with an acidifiable base.'[5] This assertion was based on Lavoisier's idea that oxygen was present in all acids—since, for example, the gas reacts with sulfur (and water) to give sulfuric acid, or with phosphorus (and water) to give phosphoric acid. On this occasion he was proved to be wrong, as he was with his assertion that the muriatic acid could be made to combine with even more oxygen to form the toxic gas first isolated by Scheele: chlorine.

Scheele's Dephlogisticated Acid of Salt

Scheele gave the first account of chlorine gas in his classic paper on manganese from 1774, where he describes how after heating the mineral pyrolusite (manganese dioxide) with muriatic acid (hydrochloric acid), 'an effervescence ensued with a smell of aqua regia'.[6] Aqua regia, the well-known mix of acids used for dissolving gold, was first prepared from mixtures of salts, but the simplest way of making it is to mix concentrated hydrochloric and nitric acids when chlorine gas can indeed be formed. Scheele, in his usual meticulous manner, goes on to prepare and study larger quantities of the yellow-green gas, enclosed in bladders or glass vessels. He describes the gas as having 'a very sensible pungent smell, highly oppressive to the lungs' and details its reactions, including how it bleaches coloured papers and flowers, and how it combines with metals—even with gold.

Scheele uses the phlogiston theory to describe how the chlorine is produced, saying that the manganese acquires 'a strong attraction for phlogiston' and removes it from the acid. He therefore refers to the chlorine gas as 'dephlogisticated acid of salt'.

Of course, the French reformers reinterpreted the formation of chlorine in terms of their new oxygen theory: rather than the hydrochloric acid losing its phlogiston, it gained oxygen to form chlorine. Fourcroy writes about the reaction in 1786 (as translated into English two years later), 'it is known that this production of a peculiar gas is due to the transition of the base of pure air, or the oxyginous principle of the calx of manganese into the muriatic acid'.[7]

Lavoisier's Oxygenated Muriatic Acid

In modern terms, we would now say that the pyrolusite (manganese dioxide) is an oxidizing agent, and that it oxidizes the negatively charged chloride ions present in hydrochloric acid to form the neutral chlorine atoms which, in pairs, make up the chlorine gas. Rather than simply adding oxygen to the hydrochloric acid, the oxygen from the pyrolusite ultimately forms water with the hydrogen from the acid, and the net result is the transferral of negatively charged electrons from the chloride ions to the manganese.

In the new nomenclature of the French, chlorine gas was called 'oxygenated muriatic acid'. Lavoisier's insistence that the gas must contain oxygen was consistent with his great theory of this substance, whose very name he intended as meaning 'acid former'. He had some experimental evidence that supported his theory that chlorine must contain oxygen. For example, when the mineral pyrolusite is heated, it gives out oxygen. If the remaining product is then reacted with hydrochloric acid, much less chlorine is formed than when the unheated pyrolusite reacts, suggesting that it was the oxygen that was crucial in forming the chlorine. Even more convincing was the observation that if a solution of chlorine gas in water is exposed to sunlight, oxygen gas is given out, and all that is left is a solution of normal hydrochloric acid. While it is tempting to conclude that the sunlight breaks down the chlorine compound into oxygen and the acid, what is crucially missed is the role of the water, and, as we now know, how it reacts with the chlorine to form hypochlorous acid, a compound similar to bleach.

So for Lavoisier and his colleagues, at the very heart of both chlorine gas and hydrogen chloride was an unknown principle. As sulfur and phosphorus combine with oxygen (and the overlooked water) to form sulfuric and phosphoric acids, so this unknown principle combined with oxygen to form muriatic acid. The French chemists named the unknown substance muriatic base or muriatic radical, 'deriving this name, after the example of Mr Bergman and Mr de Morveau, from the Latin word *muria*, which was anciently used to signify sea-salt'.[8] Even though he did not know exactly what it was, Lavoisier included this 'muriatic radical' in his list of the then-known elements.

Chlorine Named at Last

Just two years after his preparations of the alkaline earth metals, Davy published his papers on the nature of what was still then known as oxygenated muriatic acid. He tried every means possible to extract the oxygen out of both dry hydrogen chloride and dry chlorine gas. Davy found that pure charcoal (usually very good at extracting oxygen and forming its gaseous oxides), 'even when ignited to whiteness' in dry chlorine or hydrogen chloride gas, produced no change. Since chlorine gas so readily oxidizes other species, Davy was surprised by this result and wrote in 1810, 'This experiment, which I have several times repeated, led me to doubt of the existence of oxygene in that substance, which has been supposed to contain it above all others in a loose and active state.'[9] Davy resolved 'to make a more rigorous investigation than had been hitherto attempted' for the detection of the missing oxygen.

After an extensive series of experiments heating various substances, including his newly discovered metal potassium, in either hydrogen chloride or chlorine gas, Davy correctly concluded that any oxygen that is ever evolved, is formed from any water present, and 'that the idea of the existence of water in muriatic acid gas, is hypothetical, depending upon an assumption which has not yet been proved—the existence of oxygene in oxymuriatic acid gas'.[10]

Davy begins his paper with a brief history of chlorine gas, writing: 'The illustrious discoverer of the oxymuriatic acid considered it as muriatic acid free from hydrogen; and the common muriatic acid as a compound of hydrogen and oxymuriatic acid; and on this theory he denominated oxymuriatic acid dephlogisticated muriatic acid.' As we have seen in Chapter 4, hydrogen was taken by some chemists to be pure phlogiston itself. If we interpret Scheele's results in this way, dephlogisticated muriatic acid would be de-hydrogenated hydrogen chloride— that is, chlorine. Although Scheele probably did not think of his discovery in quite this way, Davy seemed to think that Scheele was essentially correct in his explanation of the formation of chlorine.

Davy does not offer a new name for the substance at this point, simply writing, 'It is needless to take up the time of this learned Society by dwelling upon the imperfection of the modern nomenclature of these substances. It is in many cases connected with false ideas of their nature and composition, and in a more advanced state of the enquiry, it will be necessary for the progress of science, that it should undergo material alterations.'[11]

However, he did give a name in his next paper on the subject, read to the Royal Society on 15 November 1810, and published the following year. In this paper, he presents numerous experiments to show that chlorine is an element with many properties similar to oxygen. He finishes with a clear argument for why the name oxymuriatic acid must be changed: 'To call a body which is not known to contain oxygene, and which cannot contain muriatic acid, oxymuriatic acid, is contrary to the principles of that nomenclature in which it is adopted; and an alteration of it seems necessary to assist the progress of discussion, and to diffuse just ideas on the subject.'[12] He then laments that Scheele did not give it a more suitable name: 'If the great discoverer of this substance had signified it by any simple name, it would have been proper to have recurred to it; but, dephlogisticated marine acid is a term which can hardly be adopted in the present advanced æra of that science.' Having learnt his lessons with names reflecting current theories, he proposes a more neutral name: 'After consulting some of the most eminent chemical philosophers in this country, it has been judged most proper to suggest a name founded upon one of its obvious and characteristic properties— its colour, and to call it *Chlorine*, or *Chloric* gas. Should it hereafter be discovered to be compound, and even to contain oxygene, this name can imply no error, and cannot necessarily require a change.' His proposed name simply derives from the Greek for 'greeny-yellow', '$\chi\lambda\omega\rho\sigma\varsigma$' ['chloros'], the colour of the pure chlorine gas.

Reluctant Acceptance

It took a few years for the conclusions from Davy's researches to be accepted. An entry from the *Memoirs of the Columbian Chemical Society of Philadelphia* from 1813 starts: 'That oxy-muriatic acid should be a simple substance appears to me as ridiculous as it is untrue.'[13] One of the reasons for this reluctance was that it meant that the theory of Lavoisier, 'the illustrious author of all that is true in chemical science',[14] that all acids contain oxygen could not be correct after all. However, as other oxygen-free acids were found, such as prussic acid (hydrogen cyanide, with the formula HCN), followed by the discovery of elements closely related to chlorine, Davy's views were gradually cemented into chemical fact.

His proposed name, chlorine, also had a rather precarious beginning. Monsieur Prieur, who translated into French Davy's paper of 1811 first proposing the name, clearly did not approve of the suggestion. He notes how Davy toyed with using a name based on the old nomenclature, which led Prieur to suggest calling oxymuriatic acid (chlorine) 'murigen' (*murigène* in French)—essentially meaning 'brine-former'. He justifies his choice by saying, 'the analogy between murigen and oxygen will be extremely noticeable. There is no one who will not immediately understand that *muride of silver, muride of tin, muride of antimony*, etc. are combinations of murigen with each of the designated metals…'[15] In his translation, Prieur added his new suggested names in every place where they should occur in Davy's paper, explaining that 'this first use will give rise to a judgment as to their propriety'. He adds 'If I have misunderstood it, these names will be dropped into oblivion, and science will have suffered very little from my attempt.' Needless to say, the names are no longer used.

The German chemists were also not immediately taken with the word 'chlorine'. In the German translation of Davy's paper, not only did translator Johann Schweigger not like 'chlorine', he also did not particularly like Prieur's alternative half-Latin-half-Greek 'murigen', which he found hard on Germanic ears. Schweigger's alternative suggestion for the word 'chlorine' was 'halogen', derived from the Greek and meaning 'salt-former'.[16] This term also did not survive for long as a substitute for 'chlorine', but eventually became used to refer to the whole group in the periodic table, since all the members readily form salts.

Eventually, however, as the other chemically related members of this group came to be discovered, Davy's recommended names came to be generally accepted. There is a lovely story of how Anna, the housekeeper of Davy's great rival, Berzelius, was once corrected by her master when, as she was washing some of his glassware, she complained of the smell of 'oxidised muriatic acid'. Berzelius replied, 'Anna, you mustn't speak of oxidised muriatic acid anymore; from now on you must say chlorine.'[17]

It was the name of chlorine that inspired the names of the halogens. The next of these to be named was the last of the non-radioactive members of the group to be isolated in its elemental form: the element fluorine.

Flowing Stones

Fluorine is named after the most abundant fluorine-containing mineral, fluorite or fluorspar. The mineral is a form of calcium fluoride, CaF_2. Agricola tells us in his *Bermannus* that the name of the mineral is derived from the Latin '*fluoere*', meaning 'to flow', since it may easily be melted in a furnace. When asked what this mineral is, his character, the master mineralogist Bermannus, replies: 'They are stones similar to gems but not so hard and called by our miners *fluores*. This is not inappropriate, I would say, since the fire melts them and makes them as fluid as ice in the sun. They are formed with a variety of pleasing colours.'[18] When asked what use fluorite has, he answers, 'During smelting it is often added to the charge for it makes it more fluid...'.

Pure calcium fluoride is pretty resistant to heat, not melting until around 1400° C. However, the naturally occurring mineral is impure and usually melts at much lower temperatures. On heating, the calcium fluoride reacts with any water present, melting and bubbling, giving out poisonous hydrogen fluoride gas and leaving the very hard-to-melt calcium oxide. Some steel producers still use fluorspar flux in their manufacturing processes to decrease the viscosity of the slag, the crust of metal oxide/silicate impurities floating on top of the molten metal.

The first person to properly study the mineral fluorite was, not surprisingly, Scheele. In his very first paper, published in 1771, he began: 'Fluor mineral is a kind of stone, especially remarkable on account of the beautiful phosphoric light which it yields in a dark place, when it has been heated. But its constituent parts are as yet little known.'[19] This property of fluorite was something we briefly touched on earlier—it was one of the earliest-discovered substances that can be made to glow. We saw that the Bolognian stone and Balduin's phosphorus were actually phosphorescent compounds—they absorb light energy, and then slowly re-emit the energy as light. We now know that fluorite glows by a different mechanism; the energy of electrons that have been excited by background radioactive processes such as high-energy cosmic rays from space can be stored semi-permanently in defects and impurities in its crystal structure. But unlike the phosphorescent compounds, which re-emit the light after a relatively short delay of up to a few hours, fluorite doesn't re-emit the energy until it is warmed up. So fluorite that has been previously 'charged' by many years' exposure to radiation seems to glow when warmed. Back in 1771, Scheele wrote, 'It is well known, that fluor mineral, after having been once thoroughly heated, loses its power of becoming phosphoric for ever after.'[20] Actually, that's not entirely true—the crystal just needs to be recharged by exposure to more radiation, but with the low levels of natural background radiation, this would take a long time. This property is used in making badges that record exposure to radiation (dosimeters), since the fluorite can be heated to essentially 'reset' it and then, after being exposed to radiation, the amount of light it gives out on being warmed will be proportional to the exact amount of exposure it received. A similar idea allows the 'thermoluminescent dating' of pottery and ceramics: their initial firing during manufacture 'resets' any such responsive minerals present, and their subsequent exposure to background radiation over the ages may then be assessed by warming and measuring how much light is emitted.

Sparry Acid

The title of Scheele's paper as it appeared in the English translation of 1786 was 'On Fluor Mineral, and its Acid'. The acid—which Scheele himself called 'fluor spar acid' ('*Fluss-spats-syra*' in his original Swedish)—is what we now call hydrofluoric acid, the very toxic acid obtained by warming the fluorite mineral (calcium fluoride) with concentrated sulfuric acid. But Scheele's work was earlier described in a text on mineralogy by John Hill in an appendix entitled 'Observations on the new-discovered Swedish Acid; and on the stone from which it is obtained'.[21] Hill states here that in addition to the term 'Swedish Acid', 'some, tho' very improperly, have given the Name of the *Sparry Acid*. Perhaps…it may be better named the *Stony Acid*; since the Substance from which we obtain it is a Stone; tho' not a Spar.' 'Stony acid' never really caught on, but the term 'sparry acid' was used in many English texts during the close of the eighteenth century, and gaseous hydrofluoric acid was sometimes referred to as 'sparry gas'. However, in their reform of chemical nomenclature, the French chemists followed Guyton de Morveau's recommendation from 1782 of avoiding 'spar' or 'sparry', since those terms are associated with many other minerals, and began to use instead 'fluor' and 'fluoric'. Salts of the acid were to be called 'fluates'—so what we now call calcium fluoride (the basis of the mineral fluorspar) was to be 'fluate of lime'.

As with his earlier thoughts on muriatic acid (hydrochloric acid), Lavoisier believed that fluoric acid was composed of an unknown element combined with oxygen. As soon as Davy had reported that muriatic acid did not contain any oxygen, proposing instead that it was made of just hydrogen and the element he named chlorine, André-Marie Ampère (1775–1836), who is now better known for his electrical discoveries, wrote to Davy pointing out similarities between the two acids. He subsequently wrote again, suggesting that that fluoric acid might also be oxygen-free and consist only of hydrogen and an as-yet-unisolated element which he would call fluorine, by analogy with Davy's chlorine.[22]

A Lucky Disaster

In Ampère's second letter, dated 25 August 1812, in which he proposes the name fluorine for the element that had yet to be isolated, he makes a casual remark that was to change the course of science forever. In the final paragraph, Ampère asks Davy if he has heard about the highly explosive oil newly discovered in Paris, composed of just nitrogen and chlorine. He warns that a violent explosion of this highly unstable compound had cost its discoverer an eye and a finger. Davy did not reply to this letter until some six months later. He then wrote: 'Sir, Till this moment I had no opportunity of replying to your obliging letter. The fulminating oil which you mentioned roused my curiosity and nearly deprived me of an eye. After some months confinement I am well again.'[23] Although he did not have time to reply to Ampère, Davy did manage to compose a paper on the new substance 'to caution the english chemists against the *oil*...' In the paper where he announces his findings, he mentions an explosion while trying to collect a considerable quantity of the oil. However, he was injured during a separate experiment to see what gases were formed during the violent decomposition of the compound. He was heating a globule of the oil under water, when suddenly 'a violent flash of light was perceived, with a sharp retort; the tube and glass were broken into small fragments, and I received a severe wound in the transparent cornea of the eye, which obliges me to make this communication by an amanuensis'. He adds, 'This experiment proves what *extreme* caution is necessary in operating on this substance, for the quantity I used was scarcely as large as a grain of mustard seed.'[24] The accident meant Davy needed assistance to write his papers. He employed a young former bookbinder who had enthusiastically attended Davy's lectures at the Royal Institution and had presented Davy with a fine bound copy of the notes he had taken (a treasure which the Institution still holds). During his subsequent time at the Royal Institution, the young lad, one Michael Faraday, became one of the most famous scientists of the nineteenth century; in addition to his chemical discoveries (which included isolating benzene for the first time), he invented the very first electric motor.

Seaweed

Before Ampère's theories about the composition of fluoric acid became widely accepted, evidence emerged that supported Davy's idea about the lack of oxygen in chlorine gas. Despite the fact that England and France were at war, Emperor Napoleon Bonaparte granted Davy, his wife, and his newly appointed young assistant Michael Faraday safe passage to Paris, where Davy was to receive a medal for his electrochemical work on the isolation of the alkali metals. During this visit, on 23 November 1813, Ampère gave Davy a strange new substance that had been discovered in Paris over a year earlier—a substance the French chemists of the day referred to 'as X, the *unknown* body'.[25]

Bernard Courtois (1777–1838), the son of a humble saltpetre manufacturer, discovered the substance by accident in 1811 while searching for the cause of corrosion in the metallic vessels used in the production of soda. As we saw in

Chapter 5, the production of saltpetre needed alkali carbonates, and these were usually obtained from the ashes of burnt plants. Rather than using the ashes from shrubs and wood, Courtois was using 'kelp'—the ashes from seaweed, or sea-wrack. Davy writes that the strange new substance 'is procured from the ashes, after the extraction of the carbonate of soda, with great facility, and merely by the action of sulphuric acid:—when the acid is concentrated, so as to produce much heat, the substance appears as a vapour of a beautiful violet colour, which condenses in crystals having the colour and the lustre of plumbago'.[26] Ampère's gift of the sample prompted Davy to investigate it immediately (using the portable laboratory he carried with him when abroad), and it was independently investigated by the French chemists. The result was four consecutive papers appearing in the journal *Annales de Chimie* between 6 and 20 December. Although Davy's experiments seemed to some of his French hosts like unwelcome interference, the mystery substance was soon recognized as a new element similar to chlorine. Furthermore, it became apparent that this new element combined readily with hydrogen to yield an acid analogous to muriatic acid—helping to cement the idea that muriatic really was, as Davy had suggested, just hydrogen combined with chlorine. The similarity to chlorine also helped in giving a name to the new substance—the French team called it *ione*, derived from the Greek for 'violet'. However, Davy proposed a slight modification for the English name:

> The name *ione* has been proposed in France for this new substance from its colour in the gaseous state, from ιον. viola; and its combination with hydrogen has been named *hydroionic acid*. The name *ione*, in English, would lead to confusion, for its compounds would be called *ionic* and *ionian*. By terming it *iodine*, from ιωδης, violaceous, this confusion will be avoided, and the name will be more analogous to chlorine and fluorine.[27]

In his paper on iodine and in his letter to Ampère, both written in 1813, Davy also refers to the element present in fluorite using Ampère's suggested name of fluorine, even though the element had not yet been isolated. But a few years later, Ampère himself seems to be having doubts about this name. As it gradually became accepted that fluoric acid did not contain any oxygen, the names that had been used for the salts of the acid became less ideal. Lavoisier and his colleagues had termed these 'fluates'. The problem was that all the names for salts that had the ending '-ate' did contain oxygen; for example, metal sulfates, nitrates, phosphates, chlorates, and iodates all contain a metal and oxygen united with sulfur, nitrogen, phosphorus, chlorine, and iodine. Metal sulfides, nitrides, phosphides, chlorides, and iodides do not contain any oxygen—just the metal and the other element. Muriatic acid and the muriate salts were both initially thought to contain oxygen, but were renamed as hydrochloric acid and chloride salts when it was shown that no oxygen was present. Rather than the name fluorine he had suggested earlier to Davy, Ampère thought 'fluore' might be suitable, whereas 'fluorure' would just be too difficult to pronounce; in the end, he thought the best name would be phtore, from the Greek meaning 'deleterious', something with the strength to ruin, destroy, or corrupt.[28] The name was used by a few chemists in the 1820s and 1830s—mainly in France. In the seventh edition of his *A System of Chemistry of Inorganic Bodies*, published in 1831, Scottish chemist Thomas Thomson wrote when introducing fluorine (which still had not been isolated): 'Ampere has given it the name of Phthorine (Phthore) from the Greek word φθοριος [phthorios], *destructive*...But it is quite evident that this new name cannot be adopted. There would be no end to names if every person at pleasure could coin new ones. The reason assigned by him for contriving this new name, namely that he was the original starter of the hypothesis, is not valid...Davy informs us that Ampere himself originally suggested the term *fluorine*.'[29]

Despite many chemists of the day, including Ampère and Davy, working hard to isolate the element fluorine, none were to succeed until decades later. Before this was to happen, another of the halogens would be isolated: bromine.

Another Muride

The isolation of iodine by Courtois caused quite a stir, and other scientists were soon repeating his experiments. Not surprisingly, this quickly led to the discovery of the related family member between chlorine and iodine in the periodic table—the element bromine. In 1825, Antoine-Jerôme Balard (1802–76), a young, recently qualified pharmacist, was investigating whether seaweed from the Mediterranean also contained iodine. In his paper *On a Particular Substance Contained in Sea-water*, published in French the following year, he described how iodine was formed on treating the lye prepared from the ashes of seaweed; its presence was easily shown by the characteristic blue colour it forms on adding a solution of starch. However, Balard added, 'there appeared, not only a blue zone of which the iodine was part of, but underneath, a zone of a rather intense yellow shade. This orange-yellow color had also appeared when I had treated the mother liquor of our saline in the same way; and the hue was all the darker as the liquid itself was more concentrated. The appearance of this shade was accompanied by a particular lively odour.' In the preliminary announcement of his discovery, read 3 July, Balard referred to his new substance as 'muride'.[30] In a more detailed report from the following month the substance is called 'brome', a name derived from the Greek meaning a stench or stink.[31] Balard later explains that the name muride, from the Latin '*muria*' meaning 'brine', had initially been recommended to him by a Monsieur Anglada. Balard writes, 'This name seemed to me eminently suited to characterize its origin, and to represent the principal circumstance of its natural history, which is connected with its discovery: it is euphonic and lends itself perfectly to the formation of the composite denominations which its combinations require.'[32] He adds that after Davy and various French researchers had 'placed chlorine among the simple bodies, a chemist proposed to call it *murigen* [*murigène*], and he reserved the name of *muride* to express the combinations of the murigen with the other simple bodies, combinations which he claimed to liken with oxides, and has since been called chlorides [chlorures]'. This chemist referred to was the translator of Davy's paper on chlorine, Prieur, whom we encountered earlier. Balard was now using the word '*muride*' in a completely different way, and it was for this reason that the commissioners appointed by the Academy to examine his work changed, with the consent of the discoverer, the name of his element into *brôme*, a name derived from the Greek for 'a bad smell'. A textbook published the following year wrote: 'This appellation may in the English language be properly converted into that of Bromine.'[33]

In the Presence of Fluorine

The similarities between the halogens chlorine, bromine, and iodine and their compounds meant chemists were confident that the similar element fluorine must exist. From the combining proportions and reactions of its compounds such as fluorite (calcium fluoride) and hydrofluoric acid, chemists could accurately predict its atomic weight. Mendeleev even included it in the very first periodic table without the element ever having been observed. The isolation of the free element proved to be no easy matter: a number of chemists (including Davy) were seriously poisoned in the attempt, and it even cost some their lives. Pure hydrogen fluoride is a volatile liquid which boils at 19.5° C, does not conduct electricity, dissolves most metals, and even dissolves glass. It is easily absorbed through the skin, causing severe, painful burns which only develop hours later. A text from 1828 states, 'The fumes of the acid must be anxiously avoided, and the hands guarded with very thick gloves, as the burns produced by the least quantity of the acid, gives the most excruciating pain, or rather tortures.'[34] The author remarks 'The fluoric acid is such a disagreeable subject to meddle with, that chemists are not fond of making experiments upon it.' Nonetheless, it was anhydrous hydrogen fluoride, with potassium hydrogen fluoride dissolved in it to make it conduct electricity, that eventually enabled Frenchman Henri Moissan (1852–1907) to isolate fluorine gas on 26 June 1886. He electrolyzed the cooled mixture in a platinum apparatus, with stoppers made of fluorite (which is not attacked by fluorine). The apparatus is shown in Figure 44.

Two days later, Moissan announced his discovery to the Academy of Science. After outlining some of the vigorous reactions of the evolved fluorine gas, including its rapid reaction with metallic mercury to form a salt, and with water to form ozone and hydrofluoric acid, he cautiously stated: 'One can, indeed, make various hypotheses about the nature of evolved gas; the simplest

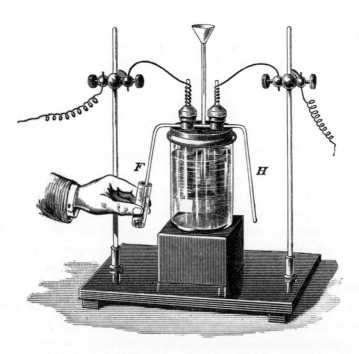

Fig. 44. An engraving from 1891 showing Moissan's apparatus to generate fluorine.

ОПЫТЪ СИСТЕМЫ ЭЛЕМЕНТОВЪ,

ОСНОВАННОЙ НА ИХЪ АТОМНОМЪ ВѢСѢ И ХИМИЧЕСКОМЪ СХОДСТВѢ.

			Ti$=$50	Zr$=$90	?$=$180.
			V$=$51	Nb$=$94	Ta$=$182.
			Cr$=$52	Mo$=$96	W$=$186.
			Mn$=$55	Rh$=$104,4	Pt$=$197,4.
			Fe$=$56	Ru$=$104,4	Ir$=$198.
		Ni$=$Co$=$59		Pl$=$106,6	Os$=$199.
H$=$1			Cu$=$63,4	Ag$=$108	Hg$=$200.
Be$=$9,4	Mg$=$24	Zn$=$65,2	Cd$=$112		
B$=$11	Al$=$27,4	?$=$68	Ur$=$116	Au$=$197?	
C$=$12	Si$=$28	?$=$70	Sn$=$118		
N$=$14	P$=$31	As$=$75	Sb$=$122	Bi$=$210?	
O$=$16	S$=$32	Se$=$79,4	Te$=$128?		
F$=$19	Cl$=$35,5	Br$=$80	I$=$127		
Li$=$7 Na$=$23	K$=$39	Rb$=$85,4	Cs$=$133	Tl$=$204.	
	Ca$=$40	Sr$=$87,6	Ba$=$137	Pb$=$207.	
	?$=$45	Ce$=$92			
	?Er$=$56	La$=$94			
	?Yt$=$60	Di$=$95			
	?In$=$75,6	Th$=$118?			

Fig. 45. Mendeleev's first published periodic table of 1869.

would be that we are in the presence of fluorine.'[35] For his remarkable achievement, Moissan was awarded the Nobel Prize in chemistry in 1906, but sadly, he died shortly after returning from the ceremony in Stockholm.

In 2012, this most reactive of all the elements—which for a long time was thought impossible to occur in nature since it reacts with almost anything it comes into contact with—was suddenly found. And it was found in fluorite, the very mineral first described by Agricola almost five hundred years earlier. A rare form of fluorite is characterized by a strange smell and is known as 'fetid fluorite', 'stinkspar', or 'antozonite'. Traces of radioactive elements such as uranium or thorium are also present in the rock, and it is the energy released from their decay that separates calcium fluoride into its elements. Tiny inclusions of fluorine gas become trapped in the fluorite—one of the few things fluorine cannot react with, since, in a sense, it has already fully reacted with it. These tiny pockets of fluorine gas are released when the mineral is crushed, explaining the fetid smell.

Ordering the Elements

Even though fluorine had not been isolated at the time, the element was included in Mendeleev's very first periodic table, drawn up in 1869 (Figure 45). In this table, the elements are ordered by atomic mass into vertical columns, and, like other groups of related elements, the members of the halogen family—fluorine (F), chlorine (Cl), bromine (Br), and iodine (I)—appear in a horizontal row. In Mendeleev's subsequent versions and in modern tables, the groups are arranged vertically. Other groups are clearly seen in the first table, such as Group 15, consisting of the elements nitrogen (N), phosphorus (P), arsenic (As), antimony (Sb), and bismuth (Bi). Famously, Mendeleev also used his table to predict the properties of elements that he thought had yet to be discovered. The first success in this field came with the filling of the space marked '? = 68'. As we shall see, Mendeleev even predicted how the element would be discovered. However, there is one entire group whose members were utterly unknown when Mendeleev first drew up his table. Remarkably, one of these elements was noticed by the eighteenth-century genius Henry Cavendish when he was studying the gases in the atmosphere. To understand the discovery of these elements, we need to turn our gaze back to the heavens.

8

FROM UNDER THE NOSE

I am a little afraid that chemists in general were piqued at being shown
that they had overlooked something which was actually under their noses...

—Travers, 1928[1]

This chapter looks at the elements in the final group of the periodic table—those elements known as the rare or noble gases. We shall see how their discovery in the atmosphere in the 1890s dates back to an observation first made by the meticulous Henry Cavendish over one hundred years earlier. This led to the unexpected discovery of an entire group of elements that needed to be added to the earliest periodic tables; and remarkably, one man was to dominate all these discoveries.

Solar Spectra

One of Isaac Newton's classic experiments was using a glass prism to split a beam of sunlight into a spectrum to show that white light is actually a mixture of all the colours of the rainbow. In 1802, William Hyde Wollaston (1766–1828), discoverer of the elements palladium and rhodium, modified the experiment by using a thin slit to admit the sunlight instead of the circular hole that Newton used. He subsequently discovered that the solar spectrum was not completely seamless, but actually contained a number of fine dark lines, now known as Fraunhofer lines. They get their name from Joseph Fraunhofer (1787–1826), who became the most skilled worker of glass and producer of lenses of the time. Using his highest-quality optical lenses, Fraunhofer observed that the solar spectrum had many dark lines; he mapped out over five hundred of these and designated the most distinct ones with the capitals letters A to H, with A and B being in the red region of the spectrum, and G and H in the violet. He used these as calibration lines in the development of better glasses for his optical instruments, and to demonstrate the superiority of his products compared with those of his competitors. The nature of the dark lines was not properly understood until the work of the German physicist Gustav Kirchhoff (1824–1997), who, in a beautiful collaboration with his colleague the chemist Robert Bunsen (1811–99), developed one of the most important analytical techniques still used in chemistry. It was with this technique that they discovered two new elements, and paved the way for others to discover many more.

The Spectroscope

It was already well known that certain substances, when put into a flame, imparted colours to it. Unlike sunlight (which produces a continuous spectrum with all the colours of the rainbow), when light from a flame coloured by a particular metal is shown through a prism, a pattern of discreet, separate coloured lines is formed. Bunsen and Kirchhoff developed an apparatus

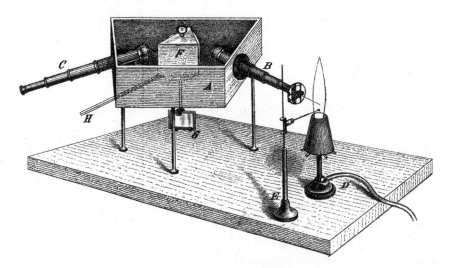

Fig. 46. The spectroscope of Bunsen and Kirchhoff. The sample to be analysed is introduced on a platinum wire (E) into the hot blue flame of the now iconic gas burner (D) that Bunsen had developed just a few years earlier. The burner has a conical shield to help steady the flame from draughts. Light from the sample enters a first telescope (B) via a narrow slit formed from two knife blades. It is then directed into a dark box housing a hollow glass prism (F) filled with a solvent to increase its refractive power. The prism is capable of being turned slightly to adjust the angles. A second telescope (C) is then used to view the separate coloured lines as the prism is turned.

known as a spectroscope to analyse the light (Figure 46) and found that each element studied (the metals from Groups 1 and 2 of the periodic table) gave its own distinctive, unique spectrum. For a given metal, the spectra seemed to be independent of the particular salt used: the bromide, chloride, hydroxide, carbonate, and sulfates of potassium all gave the same spectra and were therefore due to the presence of the potassium ions only. Similarly, the temperature of the flame or type of fuel used did not seem to alter the characteristic line spectrum of a given metal. What is more, the positions of the coloured lines seemed to match up precisely with some of the dark lines characterized by Fraunhofer. For example, the spectrum of sodium salts were dominated by two very closely spaced yellow lines exactly coincident with the dark lines labelled 'D' by Fraunhofer. These yellow lines are now known as the sodium-D lines.

As well as being able to identify the presence of different metals, the technique was also incredibly sensitive. The investigators burned a tiny 3 mg sample of sodium chlorate and sugar in one corner of the room (of about 60 cubic metres volume) while viewing the blue flame of the burner through the spectroscope in the other corner. After a couple of minutes, the telltale yellow lines of sodium could be seen through the eyepiece because of the traces of sodium ions in the smoke in the room. They calculated that they could easily detect less than one three-millionth of a milligram in weight of the salt. This was far more sensitive than any other method of analysis known at the time, and it was to revolutionize chemical analysis.

Spectral Elements

Bunsen and Kirchhoff immediately realized the potential of their new technique and started to reap its rewards. They analysed various mineral waters, minerals, and even cigar ashes, detecting trace amounts of metal salts. But their technique could go much further than detecting the previously known elements. They wrote: 'The method of spectrum-analysis may also play a no less important part as a means of detecting new elementary substances; for if bodies should exist in nature so sparingly diffused that the analytical methods hitherto applicable have not succeeded in detecting, or separating them, it is very possible that their presence may be revealed by a simple examination of the spectra produced by their flames.' They go on to give the first tentative suggestion of a new element they have discovered: 'We believe that, relying upon unmistakeable results of the spectrum-analysis, we are already justified in positively stating that, besides potassium, sodium, and lithium, the group of the alkaline metals contains a fourth member, which gives a spectrum as simple and characteristic as that of lithium—a metal which in our apparatus gives only two lines, namely a faint blue one, almost coincident with the strontium line Sr δ, and a second blue one lying a little further towards the violet end of the spectrum...'.[2] By the time of their second publication on this subject the following year, they had isolated a few grams of a salt of the new metal from 44 tonnes of the mineral water of Dürkheim. Even more impressively, they had also discovered the presence of a second new alkali metal and isolated a few grams of its salts by processing 180 kg of the mineral lepidolite.

The spectrum of their first new element was characterized by the 'two splendid blue lines situated close together'. They write: 'as no known elementary body produces two blue lines in this portion of the spectrum, we may consider the existence of this hitherto unknown alkaline element as thus placed beyond a doubt'.[3] The name for the new metal was inspired by their revolutionary technique and the appearance of the spectrum of the new element: 'The facility with which a few thousandths of a milligramme of this body may be recognized by the bright blue light of its incandescent vapour, even when mixed with large quantities of the more common alkalies, has induced us to propose for it the name *Cæsium* (and the symbol Cs), derived from the Latin "cæsius", used to designate the blue of the clear sky.'

Bunsen and Kirchhoff's second new alkali metal was obtained from the mineral Saxony lepidolite as an insoluble platinum salt contaminated with the analogous potassium salt. As they repeatedly washed the precipitate with boiling water, more of the potassium salt was removed to leave the new salt, which was again examined with their spectroscope. They write: 'two splendid violet lines, lying between the strontium line Sr δ and the blue potassium line K β will be noticed on the gradually fading continuous background of the potassium spectrum. These new lines increase in brilliancy as the washing is continued, and a number more appear in the red, yellow, and green portions of the spectrum.'[4] Once again, it was the appearance of the line spectrum that was to suggest a name for their new element. They write: 'None of these lines belong to any previously known body. Amongst them are two which are especially remarkable, as lying beyond Fraunhofer's line A and the potassium line Kα coincident with it, and therefore situated in the outermost portion of the red solar rays. Hence we propose for this new metal the name *Rubidium* (and the symbol Rb), from the Latin "rubidus", which was used to express the darkest red colour.'[5]

Green Shoots and Indigo

Bunsen and Kirchhoff presented the world with a new technique capable of detecting the most minute traces of elements, and chemists wasted no time putting it to use. In March 1861, the English chemist William Crookes (1832–1919) examined all the specimens in his extensive home laboratory with the aid of the spectroscope. Looking at some selenium residues from a sulfuric acid plant that he had had for over ten years, he was expecting to find the signature of tellurium. Instead 'suddenly a *bright green line* flashed into view and as quickly disappeared'.[6] Crookes was familiar with examining solar spectra but the 'isolated green line in this portion of the spectrum was new to me', and so he set out to discover the cause. Initially he thought the element would be similar to the elements sulfur, selenium, and tellurium, but he later recognized that it was a metal. He named his new element a couple of months after he first announced its discovery in the spring, and, like Bunsen and Kirchhoff, chose to name it from the appearance of its spectrum. He proposed for it 'the provisional name of *Thallium*, from the Greek θαλλός [*thallos*], or Latin *thallus*, a budding twig,—a word which is frequently employed to express the green tint of young vegetation; and which I have chosen as the green line which it communicates to the spectrum recalls with peculiar vividness the free colour of vegetation at the present time'.[7] In 1863, the German chemists Reich and Richter were looking for the presence of thallium in various ore samples when, instead of finding its distinctive green line, they found a new indigo-blue line. Though lacking the romantic description of Crookes, they too named the new element after its novel line in the spectrum, calling it indium.

Given the success of the spectroscope in discovering the elements caesium, rubidium, thallium, and indium, it is perhaps not surprising that Mendeleev thought that this technique would be instrumental in the discovery of some of the other elements he predicted ought to exist but were missing from his first periodic table. In 1871 he wrote an extensive article on his periodic system which included detailed predictions of a couple of the 'missing' elements. Mendeleev writes: 'The periodic law indicates the gaps which still exist in the system of the known elements, and enables us to predict the properties of the unknown elements, as well as those of their compounds.'[8] Since he wanted to avoid creating new names for undiscovered elements, Mendeleev designated them with a Sanskrit number ('*eka-*', '*dvi-*', and '*tri-*', meaning 'one-', 'two-', and 'thre'), which indicated how many places they were below known elements in the same group. Thus, for the elements directly under aluminium and silicon, 'the author has named these undiscovered elements eka-aluminium, El, and eka-silicium, Es'. Mendeleev then predicts the properties of eka-aluminium, which 'according to the periodic law should be the following:—Its atomic weight will be El = 68; its oxide, El_2O_3; its salt will present the formula ElX_3. Thus its (only?) chloride will be $ElCl_3$...and will be more volatile than $ZnCl_2$.'[9] Finally he adds, 'The volatility as well as the other properties of the saline compounds of El being the mean between those of aluminium and those of indium, it is probable that the metal in question will be discovered by means of spectrum analysis, as was the case with indium and thallium.' These predictions of Mendeleev first appeared in Russian in 1871; it seems they did not appear in English until after the element was discovered and it was realized that the predictions were impressively accurate.

Gallic Cockerels

Paul Émile Lecoq de Boisbaudran (1838–1912) started his chemical studies by carrying out experiments outside in the yard of his family home. With the financial assistance of an uncle, he fitted out a modest laboratory and there taught himself the techniques of chemical analysis. He became extremely skilled in using the spectroscope and he developed new techniques, particularly that of using an electric spark rather than the Bunsen flame to excite the electrons in the atoms of his samples. In 1874 he published a detailed work on the spectroscopic study of thirty-five elements. Suspecting there were still missing elements to be found, in February of the same year he began to investigate a 52 kg sample of the mineral blende, taken from the Pyrenees. His labours paid off, and the following year he announced: 'Between three and four in the evening of August 27, 1875, I found indications of the probable existence of a new elementary body in the products of the chemical examination of a blende from the mine of Pierrefitte.'[10] He had detected the presence of the element using his new methods. These proved to be crucial, since the Bunsen flame is not hot enough to give rise to the emission spectrum, but the spark or the much hotter hydrogen-oxygen flame is. He writes 'the few drops...in which I concentrated the new substance gave under the action of the electric spark a spectrum composed chiefly of a violet ray, narrow, readily visible, and situate at about 417 on the scale of wave-lengths. I perceived also a very faint ray at 404.' Lecoq also gave his element a name: 'The experiments executed since August 29, confirm me in the view that the body in question is a new element for which I propose the name Gallium.' Lecoq does not explain at this point why he chose the name gallium, but in a fuller report of the metal and its properties published in 1877, he says he chose the name in honour of France (Gallia). The political magazine *La Revue Politique et Littéraire* was rather skeptical of his motives, however: 'It seemed to us that patriotism had been foreign to the decision of M. Lecoq de Boisbaudran, who had simply wanted to follow the example of the scholars of the sixteenth century and Latinize his name: *the cock* in Latin is *gallus*, whence *gallium*, to designate the metal discovered by M. *Lecoq*.'[11] Lecoq's biographers state that he was anxious to point out that the metal was not named after himself. However, if he had wanted to name the element after his country and avoid such accusations he could always have called it 'francium'; this would be the choice of Marguerite Perey when she discovered the highly unstable last member of the alkali metals in 1939.

Scandium, Germanium, and Angularium

As soon as the discovery of gallium was announced, Mendeleev published his paper stating that this new element was clearly the eka-aluminium that he had predicted. This immediately brought attention to his periodic law, which became even more firmly established with the next two elements discovered that similarly agreed with his predictions. The first of these was the element scandium, found by the Swedish chemist Lars Fredrik Nilson (1840–99) in 1879 and thereafter shown to be the element Mendeleev called ekabore, or ekaboron. Nilson had been examining one of the sub-components of the complex earth yttria, obtained from the minerals gadolinite and euxenite. This was the earth first discovered in 1794, which took over one hundred years to painfully resolve into the compounds of ten new elements. Seven of these ended up with names connected with localities in Sweden: Nilson's scandium (discovered in 1879), thulium (also discovered in 1879 and named after the ancient Greek name for Scandinavia), holmium (1886, named after the Latin name for Stockholm), and finally yttrium (1843), erbium (1879), terbium (1886), and ytterbium (1907)—all, remarkably, named for the village of Ytterby on the island of Resarö, one of the many islands that make up the Stockholm archipelago.

Further confirmation of the periodic law came with the discovery of germanium six years later. This element was discovered by German chemist Clemens Winkler (1838–1904) in the summer of 1885 in a silver-rich mineral named argyrodite. A translation of the announcement mentions the difficulty encountered during the analysis of the mineral: 'However often and however carefully the analysis was conducted, a loss of 6–7 per cent always remained unaccounted for. After a long and laborious search for the source of this error, Clemens Winkler has at length succeeded in establishing the presence of a new element in argyrodite. *Germanium* (symbol Ge), as the new element is called, closely resembles antimony in its properties, but can, however, be sharply distinguished from the latter. The presence of arsenic and antimony in the minerals accompanying argyrodite, and the absence of a method of sharply separating these elements from germanium, made the discovery of the new element extremely difficult.'[12]

The chemical similarity of compounds of the new element with those of antimony caused confusion at first as to where it should be placed in the periodic table. Initially, Winkler thought he had found Mendeleev's ekastibium (eka-antimony). On 26 February 1886, immediately after reading of the announcement, Mendeleev wrote to Winkler to say that he did not think his element was ekastibium, but actually ekacadmium. Both these predicted elements, ekastibium and ekacadmium, turned out to be among a group of predictions that Mendeleev got wrong; there are no elements in these positions in the modern periodic table. Mendeleev also wrote in his letter that 'the great volatility of germanium itself and the large volatility of its chloride do not allow it to be considered as Ekasilicon, though other properties are quite close'.[13] However, it was soon established that actually the correct position was indeed to replace ekasilicon. Mendeleev's predictions were, again, impressively accurate; for example, the boiling point of germanium chloride is 87° C, and Mendeleev predicted it would be around 100° C or a little lower.

In an obituary of Winkler written by his longtime assistant Otto Brunck, we hear that Winkler initially considered the name neptunium for the element. This was because, just as the existence and precise position of the planet Neptune had been predicted by the French mathematician Urbain Jean Joseph le Verrier (who then sent the coordinates to the German astronomer Johann Gottfried Galle, who in turn observed the planet the very same night he received the correspondence), Winkler wanted to acknowledge that he had discovered the element first predicted by Mendeleev. Unfortunately, the name neptunium had been proposed around ten years earlier for another element that proved to be a false discovery, so Winkler decided on a new name. Brunck writes: 'So, on the advice of his friend Weisbach, he followed after the example of the gentlemen Lecoq de Boisbaudran and L. F. Nilson and named this the newest element after the land in whose soil it was first found "germanium".'[14] However, one of the editors of a French journal had other ideas, writing: 'It would be a fitting tribute to M. Mendeleev, and certainly due to the brilliant design of the Russian scholar, to give in the future to the elements announced by him the names he gave them himself. Let Mr Winkler begin and give the example, let him give up the name of germanium, which has a too pronounced earthy taste which can confuse it with the geranium and give his new element the name "Ekasilicium".' Brunck then questions whether the editor really thought there was no controversy about the naming of gallium, pointing out that even some Frenchmen thought that Lecoq might have named it after himself. Playing on the fact that 'Winkel' in German means 'angle', Brunck adds that 'Lothar Meyer and du Bois-Reymond jokingly advised Winkler to call the element he discovered "Angularium" in order to eliminate the pain of the French chauvinists.'

As impressive as Mendeleev's predictions for the new elements gallium, scandium, and germanium were, there was an entire group of elements that were soon to arrive on the scene which Mendeleev had not only not predicted, but which he initially resisted even accepting as elements. Evidence for the first of these new elements, and even a name for it, had actually been put forward in 1868, before Mendeleev's first table appeared. This is the only element to have first been 'discovered' off our planet—the element helium.

Emissions and Absorptions

In 1860, with the realization of the importance of the bright lines present in the flame spectra of metal salts and how they were characteristic of particular metals, Kirchhoff had finally explained the origin of the dark Fraunhofer lines in the solar spectrum. He showed that when bright sunlight passed through sodium salts in the cooler flame of an alcohol lamp, the D lines in the solar spectrum became much darker compared with the light that did not pass through the flame. In contrast, when less bright sunlight was passed through the intense yellow flame of a hotter lamp, the light emitted exactly replaced that missing from the D lines in the solar spectrum. Kirchhoff had shown the distinction between absorption and emission.

We now know that the light emitted or absorbed by a particular element is due to electrons within the atom (or ion) moving from one energy level to another. The hot flame of the Bunsen temporarily causes electrons to be promoted from their usual low energy state to a higher one. When the electrons return to their natural low state, this energy is given out in the form of light. The exact colour of the light is indicative of the precise change in energy—light at the red end of the spectrum is due to smaller energy transitions, whereas light at the blue end (or indeed ultraviolet or even X-ray light) is due to much larger energy transitions.

In contrast, if a continuous spectrum of light (representing many different energies) is passed through a gaseous sample of an element, only the colours of light which cause the precise allowed electronic energy transitions in the atoms are absorbed by the atom of the gas, and so that colour is missing from the light beyond the sample. This gives rise to a black band where the colour appears to be missing from the otherwise continuous spectrum. Kirchhoff described the spectrum as being 'reversed'—the bright lines characteristic of a particular element had become dark ones. In their classic paper from 1860, Bunsen and Kirchhoff write: 'From this we may conclude that the solar spectrum, with its dark lines, is nothing else than the reverse of the spectrum which the sun's atmosphere alone would produce.'[15] All of a sudden, an area once thought forever inaccessible—the chemical analysis of the Sun—was possible. All that needed to be done was to find the elements that when brought into a flame, produced bright lines that coincided with the dark ones of the solar spectrum. As the authors said: 'The method of spectrum-analysis not only offers, as we flatter ourselves we have shown, a mode of detecting with the greatest simplicity the presence of the smallest traces of certain elements in terrestrial matter, but it also opens out the investigation of an entirely untrodden field, stretching far beyond the limits of the earth, or even of our solar system.'[16]

Identifying the Lines

Many of Fraunhofer's lines were soon shown to be due to the presence of certain elements such as hydrogen, sodium, iron, magnesium, and calcium in the relatively cooler parts of the solar atmosphere. It also soon became apparent that some lines were due to absorptions by species in the atmosphere of the Earth rather than the Sun. These so-called telluric rays varied as the sun was observed at different times of day; the rays pass through less of the atmosphere when the sun is overhead, and so less light is absorbed compared to when the sun is on the horizon and the rays pass through much more atmosphere. Particularly valuable work in this area was carried out by Pierre Jules Cesar Janssen (1824–1907), who in 1864 made observations in the Alps at an altitude of 2700 m in order to show the absorptions decreasing, and who also observed the light from a huge bonfire (rather than from the Sun) from distances up to 21 km to demonstrate that the absorptions were due to gases in the air.

In one experiment in 1889, he observed the lamp on the recently completed Eiffel Tower from his observatory almost 8 km away: 'M. Eiffel, having very obligingly placed the Tower of the Champ de Mars at my disposal for the experiments and observations that I would like to institute there, I thought to take advantage of the powerful source of light that has just been installed there for some studies of the telluric spectrum and, in particular, that which relates to the origin of the spectral lines of oxygen in the solar spectrum.'[17] He found that two lines known as Fraunhofer A and B lines were actually due to atmospheric absorptions by oxygen molecules (O_2), and not due to species in the atmosphere of the Sun.

Perhaps his most heroic experiment was when, in his sixties and lame, he was carried on a stretcher-type contraption to the summit of Mont Blanc at an altitude of 4800 m in order to try to determine whether there were any solar absorptions attributable to oxygen, or whether they were entirely terrestrial in origin. But it is another of his observations that concerns us now. On 18 August 1868, he went to the city of Guntur near the Bay of Bengal in India to make observations with his spectroscope during a total solar eclipse. Specifically, he wanted to observe the huge jets of material occasionally ejected and then reabsorbed by the sun, known as solar prominences. During a total solar eclipse the body of the Sun is completely hidden by the Moon and so these prominences, which may extend for thousands of miles, are clearly visible, together with the so-called corona of the Sun—its extremely hot outer atmosphere, which appears as a surrounding 'crown' of light emitted by excited atoms and ions. Janssen wrote that 'immediately after the totality, two magnificent protuberances appeared' and one of them 'shone with a splendour that is hard to imagine. The analysis of its light immediately showed me that it was formed by a huge incandescent gas column, mainly composed of hydrogen gas.'[18] The spectrum of the prominence was dominated by bright spectral lines because of hydrogen (Fraunhofer's C and F lines). Janssen was surprised by how bright the light from this corona was and realized that, with carefully modified apparatus, he would be able to observe these rays even in the absence of the eclipse. Little did he know that another astronomer had already come to the same conclusion.

The Line near D

Janssen had not been the only scientist observing the solar eclipse that August. The Proceedings of the Royal Society from that period are dominated by reports from various scientific groups sent to study the event, who describe the beautiful red flames of the solar prominence. But among the papers are observations made on a normal October day in London. Inspired by the new field promised by the work of Bunsen and Kirchhoff, Norman Lockyer (1836–1920) had modified his telescope with the addition of a spectroscope. As early as 1866, he realized that while the diffuse light of the solar atmosphere weakens in intensity the more it is refracted through a series of prisms or diffraction gratings, the sharp, bright emission lines of the prominences remain unchanged and so should be easily seen without the inconvenience of waiting for an eclipse. He made his first successful observation of the solar prominence on 20 October 1868 and immediately sent news of his observations to the Secretary of the Royal Society, who received it the following day:

> October 20, 1868
> Sir,—I beg to anticipate a more detailed communication by informing you that, after a number of failures, which made the attempt seem hopeless, I have this morning perfectly succeeded in obtaining and observing part of the spectrum of a solar prominence.
>
> As a result I have established the existence of three bright lines in the following positions :—
> I. Absolutely coincident with C.
> II. Nearly coincident with F.
> III. Near D.
> The third line (the one near D) is more refrangible than the more refrangible of the two darkest lines by eight or nine degrees of Kirchhoff's scale. I cannot speak with exactness, as this part of the spectrum requires remapping.
>
> I have evidence that the prominence was a very fine one.
> The instrument employed is the solar spectroscope, the funds for the construction of which were supplied by the Government-Grant Committee. It is to be regretted that its construction has been so long delayed.
> I have &c., J. Norman Lockyer.[19]

The importance of this announcement (other than the groundbreaking fact that it was made without the assistance of a total eclipse) was in the phrase 'Near D'. Janssen and other observers had made almost the same observations during the eclipse, but, if they made any such distinction at all, they had thought that the orange-yellow line was *at* D rather than *near* D, and so was likely connected with the famous sodium D lines. With their new, modified apparatus, both Janssen and Lockyer were able to observe simultaneously the emission lines from the solar flares, and the usual continuous solar spectrum

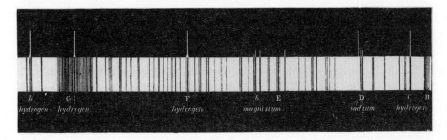

Fig. 47. One of Lockyer's photographs showing the lines of the emission spectrum superimposed on top of the absorption spectrum with its black Fraunhofer lines. Directly above the black D lines in the absorption spectrum are two short lines in the emission spectrum owing to sodium atoms. Immediately to the left of these lines is a tall line with no coincident dark Fraunhofer line beneath. This line is labelled with a question mark.

with its dark Fraunhofer lines. Comparing the two revealed that the while the bright lines at C and F attributable to hydrogen were at the same position as the dark Fraunhofer lines, the orange-yellow line 'near D' was not at exactly the same position. One of Lockyer's later improved images shows the bright emission lines from the prominence as white lines positioned above the usual solar spectrum with its dark absorption lines (Figure 47). Most of the emission lines coincide exactly with the dark absorption bands in the solar spectrum, and are marked with the names of the elements that cause them. However, immediately to the left of the sodium D lines (with small emission lines above them) is a taller emission line with no corresponding dark line. This is simply labelled with a question mark.

The Imaginary Substance Helium

Being located near the two sodium D lines, the new line soon became known as the D_3 line. Lockyer was hesitant to announce immediately that his new line must be due to a new element. Instead he sought out the advice of a chemist more experienced with laboratory spectroscopy, Edward Frankland (1825–99), in order to see if the line might be due to hydrogen under more extreme conditions of pressure and temperature. After extensive experiments it seems that Lockyer believed the line was due to a new, undiscovered element, but Frankland was not convinced. Curiously, it seems Lockyer never directly referred to his spectral element by name (at least in print) even though he had plenty of opportunity of doing so as editor of the journal *Nature*, which he had founded in November 1869. However, other workers in the field did know that he thought a new element was the likely cause for the D_3 line and that he referred to it as 'helium', from the Greek word for the Sun, '*helios*'. The first appearance in print seems to be in the report of the inaugural address of the new incoming president of the British Association for the Advancement of Science, given on 2 August 1871 and published in full in *Nature* the following day. In it, the new president, Sir William Thomson, gives an account of the current state of science. When referring to the recent progress in solar spectroscopy, he mentions how 'the chemist and the astronomer have joined their forces' and how 'an astronomical observatory has now, appended to it, a stock of reagents such as hitherto was only to be found in the chemical laboratory'.[20] He later refers to the eclipse: 'During six or eight precious minutes of time, spectroscopes have been applied to the solar atmosphere and to the corona seen round the dark disc of the moon eclipsing the sun.' He adds: 'It seems to have been proved that at least some sensible part of the light of the "corona" is a terrestrial atmospheric halo or dispersive reflection of the light of the glowing hydrogen and "helium" round the sun.' A footnote to the report says 'Frankland and Lockyer find the yellow prominences to give a very decided bright line not far from D, but hitherto not identified with any terrestrial flame. It seems to indicate a new substance, which they propose to call Helium.'

Perhaps to counter this over-confident remark, the incoming president of the same society for the following year, William B. Carpenter, gave a rather different interpretation in his address: 'Mr Lockyer speaks as confidently of the Sun's Chromosphere of incandescent Hydrogen, and of the local outbursts which cause it to send forth projections tens of thousands of miles high, as if he had been able to capture a flask of this gas, and had generated water by causing it to unite with oxygen. Yet this confidence is entirely based on the assumption that a certain line which is seen in the Spectrum of a hydrogen flame *means* hydrogen also when seen in the spectrum of the Sun's chromosphere.' And as for the new element: 'But when Frankland and Lockyer, seeing in the spectrum of the yellow Solar prominences a certain bright line not identifiable with that of any known Terrestrial flame, attribute this to a hypothetical new substance which they propose to call Helium, it is obvious that their assumption rests on a far less secure foundation; until it shall have received that verification, which, in the case of Mr Crookes's researches on Thallium, was afforded by the actual discovery of the new metal, whose presence had been indicated by him by a line in the Spectrum not attributable to any substance then known.'[21] Given that Lockyer does not seem to have explicitly claimed the existence of a new element (at least not in writing), this criticism seems rather unfair. However, it was not an isolated case.

Mendeleev also seems to have been incredulous when it came to this new element. Speaking at the Royal Institution in London in 1889, he remarked on 'the imaginary substance *helium*' and concluded, 'all probabilities are in favour of the helium line simply belonging to some long since known element placed under such conditions of temperature, pressure, and gravity as have not yet been realised in our experiments'.[22] Mendeleev thought there was no place in his periodic table for helium; but eventually he had to make room, not just for this one element, but all its relatives too in the form of a whole new chemical group. The story starts with the density of nitrogen.

The Nitrogen Problem

In the first periodic tables, the elements were arranged by their atomic weight. We now know that the correct order is actually by atomic number, which is the number of positively charged protons present in the nucleus of an atom of the element, and is unique for each element. The structure of the nucleus was not known in the nineteenth century, but since the order of the elements by atomic weight is almost the same as when they are ordered by atomic number, it was still possible to construct meaningful tables. Nonetheless, much attention was paid to the precise values of atomic weights. One theory proposed that the heavier elements are essentially made up of multiples of hydrogen atoms: the atomic weight of carbon being twelve times that of hydrogen; the atomic weight of oxygen sixteen times. But were these weights *precisely* integer multiples (which would support the theory), or were they just *approximately* twelve and sixteen times? In order to answer these questions, the atomic weights needed to be measured extremely accurately. The task of measuring those of the gases hydrogen, oxygen, and nitrogen was undertaken by John William Strutt, who on the death of his father in 1873 inherited the Barony of Rayleigh and became Lord Rayleigh.

Rayleigh's method was essentially to weigh a glass flask when evacuated, and then to compare it with the weight of the flask filled with the pure gas. But this brief description gives no indication of the years of arduous labour the process actually involved. For a start, the specially constructed balance used for the weighing needed to be in its own temperature- and pressure-controlled room. Before weighing, the balance needed to settle overnight, and then it had to be read from outside the room through a window so as not to disturb it inside. Rayleigh's measurements were so precise that he could even detect the fact that the glass flask was ever so slightly smaller when it was evacuated, since the pressure of the atmosphere pressing on it from outside was no longer balanced by pressure from inside. This subtly altered the buoyancy of the flask in air, making a small but measurable difference. Trying to prepare the gases in as pure a way as possible was another problem—but all of these obstacles were eventually overcome. After more than three years of work, Rayleigh found that the atomic weight of oxygen was 15.912 times that of hydrogen—not an exact integer value. After working on oxygen and hydrogen, Rayleigh turned to what he thought would be the simpler problem of nitrogen.

Rayleigh prepared his nitrogen by removing the oxygen, water vapour, and carbon dioxide from the air, and did indeed obtain a value quite quickly. However, he thought he should repeat the measurement with nitrogen prepared using a different method. This time, he bubbled air through concentrated ammonia solution to produce a mixture of ammonia (NH_3) and air. This mixture was passed over a hot copper catalyst, which allowed the oxygen and ammonia to react to form nitrogen and water. The water and any surplus oxygen were removed to give nitrogen that had come partly from the air, and partly from the ammonia. But now Rayleigh did not get the same result as before: the atomic weight was ever so slightly lower. After checking and rechecking, he always found a discrepancy which could not be accounted for.

Rayleigh was so puzzled that he wrote a letter to the journal *Nature* in 1892. It began:

> I am much puzzled by some recent results as to the density of *nitrogen*, and shall be obliged if any of your chemical readers can offer suggestions as to the cause. According to two methods of preparation I obtain quite distinct values. The relative difference, amounting to about 1/1000 part, is small in itself; but it lies entirely outside the errors of experiment, and can only be attributed to a variation in the character of the gas...The question is, to what is the discrepancy due?[23]

Rayleigh makes some obvious suggestions, such as known impurities, but then explains why these can be ruled out. Eighteen months later, he presented a more detailed paper in which he now compared the density of nitrogen prepared from the air with pure chemically prepared nitrogen and found the difference to be even larger—about 0.5 per cent.

Around this time, the chemist William Ramsay (1852–1916), who had been in earlier correspondence with Rayleigh, decided to see if he could help to solve the puzzle. The two scientists approached the problem in different ways. Ramsay took nitrogen prepared from the air and passed it over heated magnesium metal to form solid magnesium nitride. During this process, he found the residual gas increased in density. In contrast, Rayleigh used the method of Cavendish from over one hundred years earlier—he sparked nitrogen with oxygen to form nitrogen oxides, which were then removed by dissolving in alkali. Both methods yielded a new gas. On 4 August 1894, Ramsay wrote to Rayleigh: 'Dear Lord Rayleigh,—I have isolated the gas at last. Its density is 19.075, and it is not absorbed by magnesium.' He goes on to give a few more details, and finishes, 'I should much like to talk to you about this. Are you going to be at Oxford? If so we will meet there. I didn't want to trespass on your preserves and yet I feel that I have done so.'[24] (Oxford was the venue for a forthcoming meeting of the British Association for the Advancement of Science.) Immediately on receiving the letter, Rayleigh replies: 'Dear Prof. Ramsay,—I believe that I too have isolated the gas, though in miserably small quantities. When I spark away (after Cavendish) 50 c.cs. of air with oxygen added as required, I get a residue of .3 c.c. which is neither oxygen nor nitrogen (no hydrogen).'[25] Rayleigh refers to the new gas as 'X' and proposes that they give a preliminary announcement at the meeting in Oxford and prepare a joint publication. Ramsay replies: 'I think that joint publication would be the best course, and I am much obliged to you for suggesting it, for I feel that a lucky chance has made me able to get Q in quantity (there are two other X's, so let us call it Q or Quid).'[26]

The announcement of the new discovery was made at the meeting of the British Association in Oxford on 13 August 1894. There were no official publications at the time, since Ramsay was keen for the full report to be submitted to the Smithsonian Institution of Washington in competition for a prize of $10,000 'for a treatise embodying some new and important discovery in regard to the nature and properties of atmospheric air'. *The Times* reported the following day:

Yesterday was a busy day in the nine sections, and one of them produced a surprise. It had been whispered about on Saturday and Sunday that an interesting announcement would be made at the meeting of the Chemical Section, and the large theatre of the Museum was well filled at half-past 10 o'clock. It appears that Lord Rayleigh has been working on the densities of gases, and he has found that nitrogen derived from the atmosphere is a little heavier than nitrogen derived from various other sources. His lordship could not explain the fact, and he published it for the information of others. Professor Ramsay took up the matter in the belief that the nitrogen derived from the atmosphere could not be pure, and he found that there is in the atmosphere a small quantity of a gas which is still more inert than nitrogen.[27]

The new gas was described as making up 1 per cent of the atmosphere, with a spectrum with 'a single blue line much more intense than a corresponding line in the nitrogen spectrum'.

Aeron, Mrs Harris, and Oxfordgen

Rayleigh remarked in a letter to Lady Frances Balfour, a driving force in the women's suffrage movement:

> The new gas has been leading me a life. I had only about a quarter of a thimble-ful, and that was not much to go upon. To get larger quantities I had to set up a dynamo, and work it for days. I now have a more decent quantity, but it has cost about 1,000 times its weight in gold! It has not yet been christened. One pundit whom we consulted suggested äeron, but when I have tried the effect privately, the answer has usually been, 'When may we expect Moses?'[28]

Lady Rayleigh was told by Lord Halsbury 'in a spirit of banter that he understood the gas was known as "Mrs Harris"'.[29] This was a reference to an imaginary friend of Mrs Gamp, one of the characters from Charles Dickens's novel *Martin Chuzzlewit*.

Another name proposed in jest appeared in the satirical French publication *Le Journal Amusant*, 25 August, 1894, which reported that:

> The English have put a proper spoke in the wheel since they are about to come first in a great chemical race. A professor at Oxford University has discovered a new gas of the air. Until now, we were content with just oxygen and azote [nitrogen] in the air. The Oxford chemist invents a third. There is great excitement in all laboratories of the five continents. The name of the new gas is not yet designated, but we think that by baptizing it *oxfordgen* we will have helped find a solution that is needed. There is an ox in our proposition, and, moreover, a delicate reminder of the city, now glorious, which gasifies us [...] When we think that humanity could live for thousands of years without knowing that the air contained different gases, we are really astounded. For a hundred years,

thank God, we could learn that it contained two. And here, five years before the coming century, we have discovered a third! Don't think that it will stop in such a good way. Chemistry never stops. In 1950, a fourth gas of the air will be discovered. In 1990, it will be the turn of a fifth. And so on, at the rate of two gases per century, at least until the end of the world...

In fact they were not too far off with their predictions; just a little out on the dates, and the fact that neither Ramsay nor Rayleigh were actually professors at Oxford.

As it happens, the name eventually chosen for the new gas was first suggested during the Oxford meeting. In his biography of his father, Robert Strutt, the fourth Baron Rayleigh, writes: 'I was not present, but so far as I can remember to have heard, no comments of much significance were made beyond civil remarks by the Chairman, and a suggestion of Mr H. G. Madan that the gas should be called argon (Greek αργον, idle) on account of its chemical inertness. This suggestion was ultimately adopted.' He adds in a footnote that 'The word occurs in the New Testament in the parable of the labourers in the vineyard, some of whom "stood idle in the market place".'[30] The name was not used straight away, since it was felt that first the gas really should be shown to be truly inert. It was not until three months after the Oxford meeting that Ramsay wrote to Rayleigh saying, 'Seeing that X is very inactive, what do you think of argon, α-εργον for a name?' To date, no stable compounds of argon chemically combined with other elements have been made, fully justifying its name of 'in active'.

Argon at Last

Many scientists seemed initially reluctant to accept the new discovery. Following its announcement in *The Times*, James Dewar (1842–1923), who had been working at the Royal Institution on the liquefaction of air and its gases and is now perhaps best known for his invention of the vacuum flask, wrote a letter to the editor of that newspaper to say that no sign of the new gas had been seen in his experiments. This was followed by a further letter from him on 18 August to say that, in the same way that ozone molecules, O_3, may be produced by passing an electrical discharge through oxygen, perhaps the new gas is a form of nitrogen—most likely an extremely inert form of nitrogen with the formula N_3. This idea also seemed to be favoured for some time by Mendeleev, who was reluctant to find a space for argon in his periodic system.

Chemists were getting restless to hear a full report of the gas. At a meeting of the Chemical Society on 6 December 1894, Dewar gave an account of his work on liquid air, and again cast doubt on the existence of a new element in the air. Sadly, both Ramsay and Rayleigh declined to attend the meeting. A report of the meeting that appeared in *The Chemical News* remarked, 'It was useless to deny that special interest attached to the communication to which they had just listened, but, unfortunately, in the absence of Lord Rayleigh and Professor Ramsay, they were left in the position of having to play "Hamlet" with only the ghost present, and, under such circumstances, the play obviously could not be continued to a successful issue. Chemists were deeply interested by the statements relating to the discovery of a new constituent of the atmosphere, brought before the British Association at Oxford, but they awaited further information before making up their minds.'[31]

Finally, the 21 December issue of *The Chemical News* reported, 'At the last meeting of the Royal Society it was announced that a paper by Lord Rayleigh and Professor Ramsay on the New Gas, to which the name of Argon has been provisionally given, will be taken as the subject of discussion at the meeting on January 31st, 1895.' It was added in a footnote that Argon was derived from the Greek '$\alpha\nu$-$\epsilon\rho\gamma o\nu$; *no work*. Symbol, A'.[32] Years later, this was changed to Ar to be more in line with the symbols from the other elements from the group that were to join it.

Exciting though the idea of a hitherto unknown element in our atmosphere must have been, colourless gases do not make for a public spectacle. Ramsay exhibited a specimen of argon during the Royal Society lecture in January, and Rayleigh's son recalls:

> I remember that Ramsay said he had been asked by friends to show some argon, and he produced a sealed glass tube to satisfy them, though of course there was nothing to be seen. Rayleigh said afterwards, 'I did not know you had as much as that.' 'I did not say what pressure it was at,' replied Ramsay. 'I was not going to risk losing a valuable stock by the tube being broken!' It was a tube from which nearly all the argon had been pumped out before it was sealed up![33]

Cavendish's Argon

Rayleigh removed the nitrogen from air following Cavendish's method, in which nitrogen and oxygen were made to combine together with the aid of an electric spark. In their joint paper, Ramsay and Rayleigh praise their predecessor's technical skill: 'Attempts to repeat Cavendish's experiment in Cavendish's manner have only increased the admiration with which we regard this wonderful investigation. Working on almost microscopical quantities of material, and by operations extending over days and weeks, he thus established one of the most important facts in chemistry.'[34] In addition to showing that nitrogen can be made to react with oxygen, Cavendish wanted to establish whether all that was left after oxygen was removed from the air was nitrogen, or if there was something else present. In the process, he undoubtedly isolated argon in 1785. He even estimated that any remaining constituent (beyond nitrogen and oxygen) is present in the atmosphere to the extent of about 1/120; the modern figure is 1/107. However, Cavendish is not usually credited as the discoverer of the gas, since he did not recognize it as a new element nor fully characterize it, and also since this side of his work was essentially overlooked for more than a century.

Anglium, Scotium, and Hibernium

Ramsay was one of the earlier supporters of the periodic table in the UK, and in 1891 he wrote a textbook based on it—A System of Inorganic Chemistry—since, as he states in the preface, 'Nearly twenty-five years have elapsed since the discovery by Newlands, Mendeleeff, and Meyer of the periodic arrangement of the elements; and, in spite of the obvious guide to a similar classification which it furnishes, no systematic text-book has been written in English with the periodic arrangement of the elements as a basis.' The periodic table that he included in his book is shown in Figure 48. Of course, when he wrote the book, his investigation of the gases of the atmosphere had not yet started. As in all versions of the periodic table around this time, helium did not feature, nor was it even mentioned in the book. The table Ramsay used may look a little odd to chemists familiar with the modern form since, as was common at the time, each of the groups labelled I–VII actually contain elements from what are now usually shown as two separate groups. There are some similarities between the elements of two 'sub-groups'—for example, in Group VI, as sulfur forms sulfuric acid, H_2SO_4, so chromium forms chromic acid, H_2CrO_4. The elements in Group VIII appear in three separate groups in modern tables.

As soon as it looked like there was a new element present in the air, Ramsay was keen to think where it might fit into Mendeleev's periodic table. Before he had even isolated it, as he was enriching the proportion of argon in his sample of

The Elements, arranged in the Periodic System.

I (a)	I (b)	II (a)	II (b)	III (a)	III (b)	IV (a)	IV (b)	V (a)	V (b)	VI (a)	VI (b)	VII (a)	VII (b)	VIII
	1 H													
Li 7		Be 9		B 11		C 12		N 14		O 16		F 19		
	23 Na		24·5 Mg		27 Al		28·5 Si		31 P		32 S		35·5 Cl	
K 39		Ca 40		Sc 44		Ti 48		V 51·5		Cr 52·5		Mn 55		Fe 56, Co 58·5, Ni 58·5.
	(63·5 Cu)		65·5 Zn		70 Ga		72·5 Ge		75 As		79 Se		80 Br	
Rb 85·5		Sr 87·5		Y* 89		Zr 90		Nb 94		Mo 95·5		? 100		Ru 101·5, Rh 103, Pd 106·5.
	(108 Ag)		112 Cd		114 In		119 Sn		120·5 Sb		125 Te		127 I	
Cs 133		Ba 137		La* 142·5		Ce 140·5		?‡ 141		?§ 143		?‖ 150		? 152, ? 153, ? 154.
	156 ?		158 ?		159 ?		162† ?		166 ?¶		167 ?		169 ?	
? 170		? 172		Yb* 173		? 177		Ta 182·5		W 184		? 190		Os 191·5, Ir, 193, Pt 194·5.
	(197 Au)		200 Hg		204 Tl		207 Pb		208 Bi		214 ?		219 ?	
? 221		? 225		? 230		Th 232·5		? 237		U 240		? 244	

* Position doubtful.   † Terbium?   ‡ Neodymium?   § Praseodymium?   ‖ Samarium?   ¶ Erbium?

NOTE.—The atomic weights are in this table given only to the nearest half unit.

Fig. 48. The periodic table, from Ramsay's *A System of Inorganic Chemistry*, 1891.

nitrogen and noting the rising density, Ramsay did a few calculations to establish what the density of the pure gas might be. His calculation was quite close, but he initially thought that the gas would, like almost every other known gaseous element, consist of two atoms united to form a molecule, as with gaseous oxygen, nitrogen, hydrogen, and chlorine, which exist as molecules with the formula O_2, N_2, H_2, and Cl_2. If this were the case, the atomic weight of the new element would be around 20. A letter from Ramsay to Rayleigh dated 24 May 1894 ends: 'Has it occurred to you that there is room for gaseous elements at the end of the first column in the periodic table?'[35] He then includes a quick sketch of a periodic table and places 'X X X' in Group VIII immediately after fluorine (F) and above the metals iron (Fe), cobalt (Co), and nickel (Ni). It seems he thought that there might be three elements with similar atomic weights around 20, and that the new element in air might be one of these, or a mixture of them. He later gave an account of these thoughts, together with suggested names for the three elements: 'The discovery of argon at once raised the curiosity of Lord Rayleigh and myself as to its position in this table. With a density of nearly 20 [relative to O_2 having a density of 16], if a diatomic gas, like oxygen and nitrogen, it would follow fluorine in the periodic table; and our first idea was that argon was probably a mixture of three gases, all of which possessed nearly the same atomic weights, like iron, cobalt, and nickel. Indeed, their names were suggested, on the supposition, with patriotic bias, as Anglium, Scotium, and Hibernium!'[36] So—following the trend of Lecoq, Nilson, and Winkler, who had chosen to honour their homelands France, Scandinavia, and Germany—Rayleigh and Ramsay thought of names based on the Latin versions of 'England', 'Scotland', and 'Ireland'.

When they isolated enough argon to study its properties, the gas was surprisingly found to exist as individual atoms, not paired up to form molecules. (Gaseous mercury was also known to exist in this state, so the finding was not entirely unprecedented.) This meant the atomic weight of the element was now around 40, and that it needed to follow chlorine in the table. There were still problems with this placement because the elements were ordered by their atomic weights rather than their atomic number. Although the resultant orders are quite similar, there are a few places in the periodic table where they differ. One example is the order of cobalt and nickel, and this was the reason for initially suspecting that these elements might contain small impurities of another metal which gave rise to the false element known as 'gnomium' that we met in Chapter 2. In a similar manner, argon (atomic number 18) has a relative atomic mass of 39.948, whereas for potassium (atomic number 19) it is 39.098. Before the real reason for the ordering was understood, Ramsay thought that perhaps the argon he had isolated might contain impurities of a heavier element with similar properties.

More Gases

Ramsay's next major discovery was prompted by a hint from the assistant keeper of the mineral department of the British Museum, Henry Miers. Miers had not been able to attend the announcement of argon on 31 January, but he wrote to Ramsay the following day: 'I do not know whether you mentioned yesterday uranium as an element with which you had experimented in connection with nitrogen and argon. The frequent presence of nitrogen (?) in the natural uranates...suggests that it might be worth while to experiment in this direction; probably you have already done so, and in that case you will pardon me for calling your attention to Hillebrand's results.'[37] Miers was referring to a paper from the American geologist William Hillebrand, who had been analysing various minerals of uranium and found, much to his (and everyone's) surprise, that they seemed to contain nitrogen gas. Hillebrand writes: 'The most surprising discovery, however, was that nitrogen is an integral component of most uraninites and possibly of all, in quantities ranging from mere traces up to over 2.5 per cent. The nitrogen is set free from the mineral as nitrogen gas by the action of non-oxydizing inorganic acid...'[38] This prompted Ramsay to try and secure some of the uranium minerals, and Miers sent him a list of dealers. In a reply to a later letter, Ramsay seems to think he might at last find a compound of argon combined with another element: 'I hope you have put me on the track of a compound. I have always held that if argon formed a compound it must be with some rare element. It could have been discovered years ago if it had formed one with any of the common elements. But it was an almost hopeless task to begin on *all* the rare elements, though I had made up my mind to try.'[39]

Crypton Revealed

Ramsay set one of his students the task of extracting the gas from samples of the mineral cleveite. After cleaning up the sample, they observed its spectrum and were in for a surprise. Ramsay wrote to his wife on Thursday, 14 March 1895:

'I have another new gas, I think, from the mineral clèveite. There is very little of it, but it isn't nitrogen, and it isn't argon. It has a very distinct but quite different spectrum. This is the brother [illegible]. We shall see.' Still thinking that argon ought to be in Group VIII of the periodic table above the iron-cobalt-nickel triad with two similar elements, Ramsay's first thought when the spectrum of his gas from cleveite promised a new element was that it might be one of these missing 'brothers' of argon. Ramsay called his unknown gas 'krypton', derived from the Greek word meaning 'hidden'. By the following Sunday, Ramsay had got his friend William Crookes, the spectroscopist who discovered thallium, to look at the gas. In a letter from that day he writes: 'Crookes thinks its spectrum is new; and I don't see from the method of treatment how it can be anything old, except argon, and that it certainly isn't. We are making some more of it, and in a few days I hope I shall have enough of it to do a density. I suppose it is the sought-for "krypton", an element which should accompany argon.'[40]

On Sunday 24 March, Ramsay wrote to his wife: 'Let's take the biggest bit of news first. On Friday I bottled the new gas in a vacuum tube, and arranged that I could see its spectrum, and that of argon, in the same spectroscope, at the same time. There is argon in the gas; but there was a magnificent yellow line, brilliantly bright, not coincident with, but very close to the sodium yellow line. I was puzzled, but began to smell a rat. I told Crookes; and on Saturday morning, when Harley and Shields and I were looking at the spectrum in the dark room, a telegram came from Crookes.' Ramsay included a copy of the historic telegram from Crookes, which simply said: 'Crypton is Helium, 58749. Come and see it.' The number referred to here is the wavelength of the troublesome D_3 line that we encountered earlier, 587.49 nm. The title of Ramsay's paper, sent to the Royal Society on 26 March 1895, was 'On a Gas Showing the Spectrum of Helium, the Reputed Cause of D_3, One of the Lines in the Coronal Spectrum'. Finally, helium, the almost mythical element first detected in the Sun by Janssen and Lockyer in 1868, had now been found on Earth, and Ramsay immediately sent a sample and note to Lockyer to inform him of the discovery. We now know that the element appears in minerals of uranium and other radioactive elements, since it is formed during radioactive decay. So-called alpha particles are actually positively charged helium nuclei which, on grabbing electrons, form neutral atoms of helium that become trapped in the rocks. This is why the helium gas which we use to fill our party balloons actually comes from the ground—it is sometimes found in deposits of natural gas where, after having been formed during radioactive decay, it becomes trapped with the other gases.

An Undiscovered Gas

Ramsay was still troubled about the position of his two new elements in the periodic table. Now that helium had been discovered, he began to think there might be a whole new group to add. In August 1897 he gave an address to the chemical section of the British Association, who were meeting that year in Toronto, Canada. This was the report in which Ramsay confessed his earlier thoughts on Anglium, Scotium, and Hibernium. However, now that helium had been discovered, he strongly suspected that there would be more gases to come: 'There should, therefore, be an undiscovered element between helium and argon, with an atomic weight 16 units higher than that of helium, and 20 units lower than that of argon, namely 20 ... And pushing the analogy still further, it is to be expected that this element should be as indifferent to union with other elements as the two allied elements.'[41] He added that the search was underway: 'My assistant, Mr Morris Travers, has indefatigably aided me in a search for this unknown gas. There is a proverb about looking for a needle in a haystack; modern science, with the aid of suitable magnetic appliances, would, if the reward were sufficient, make short work of that proverbial needle. But here is a supposed gas, endowed no doubt with negative properties, and the whole world to find it in. Still, the attempt had to be made.'

The clue as to where to look for the needle came from the Irish physicist George Johnstone Stoney, who had calculated the likelihood for gases of different mass of escaping Earth's gravity and leaving the atmosphere. He calculated that something with a relative atomic mass of 20 ought to remain in the atmosphere. On hearing this, Ramsay realized that once oxygen and nitrogen had been removed from the air by the usual chemical means, the gas that was left might contain the elusive new gas as a tiny impurity in what was essentially argon. This proved to be the case.

The technical breakthrough that enabled the discovery was a method patented in early 1898 by Dr William Hampson for producing relatively large quantities of liquid air. The prototype apparatus was immediately acquired by the company that would later become the British Oxygen Company, and Hampson kindly brought to Ramsay samples that had been produced 'out of hours'; the first of these, 750 ml of liquid air, arrived on 24 May 1898. Unsure what to do with it, Ramsay and his assistant, Morris Travers, did what all students do when first encountering liquid nitrogen—they played about, freezing rubber, etc. But the most important thing they did was to allow it to slowly evaporate and then trap all the gas from the last fraction. Not suspecting that it would yield any result at all, they did not process the gas in the usual way until a week later—removing the oxygen over hot copper, and then the nitrogen over hot magnesium. They then added more oxygen to spark the remaining mixture over lunch in order to remove the final traces of nitrogen. On returning from lunch a colleague joked with Travers, saying, 'New gas this afternoon, Travers?' 'Sure thing,' replied Travers with a laugh. He later described the next step: 'The removal of the oxygen was carried out as usual, and a little of the gas was introduced into a spectrum tube. On examination with the direct-vision prism it was at once obvious that though the spectrum showed the characteristic argon lines distinctly, it was dominated by a very bright green line on the yellow side of the mercury green line, and by a very bright yellow line with a greenish tinge, and definitely not one of the known D lines. The tube certainly contained argon, and also, and with equal certainty, a hitherto unknown gas.'[42]

It was only later that they realized that if they were looking for the element lighter than argon, they should have taken the very first fractions boiling off the liquid air, not the last fractions, which would contain any heavier gases. Still, since they had discovered a new gas, it hardly mattered.

Ramsay wrote to Rayleigh to tell him of the discovery: 'You are the first person, outside the family, to whom I write to tell that Travers and I have succeeded in isolating a new gas from air...I have sent a note to the Royal Society, and also to the Academie des Sciences. Thanks to Hampson, we can get any quantity...We have about 15 litres of argon ready for a hunt for the lighter constituent. I think we shall hunt it up next, before going on with this one. I have taken the name, krypton, for it; you remember, we considered it, and rejected it for argon.'[43]

The first public announcement of the gas in English appeared in The Times on Tuesday 7 June 1898: 'A NEW GAS. (from our own correspondent) Paris, June 6. At to-day's sitting of the Academy of Sciences M. Berthelot read a letter from Professor Ramsay, the co-discoverer with Lord Rayleigh of argon, giving the first announcement of another discovery of the same nature. This new gas he proposes to call crypton. The discovery was effected, like that of argon, by the aid of the spectroscope...'.[44]

The New Light Gas

On the very same day that the announcement of krypton appeared in The Times, work began on the search for the lightest component. Dr Hampson brought some more liquid nitrogen, and Travers and Ramsay cooled some atmospheric nitrogen gas and then collected the first gas that came off as the liquid re-evaporated. The spectrum looked promising, with new lines present in the violet, red, and green. Although the results were not yet conclusive enough for a publication, the new element received its name at this stage. The story is recounted by Ramsay's assistant, Travers, who describes how Ramsay's young son, Willie, 'had come to the laboratory to see how the new gas, krypton, was made, and at the moment we thought that we had an indication of lighter gas in the first boilings from liquefied atmospheric nitrogen.' On being shown the newer spectrum, the lad asked: '"What are you going to call it? I should call it Novum." His father replied "I think that we had better go to the Greek, and call it Neon."'[45] And so the latest gas was named after the Greek (rather than the Latin) for 'new', following the suggestion of a thirteen year old.

During the presentation of the paper on krypton at the Royal Society, Ramsay had heard that Dewar would soon be commenting about a lighter gas in the atmosphere—presumably the neon that Ramsay's team had also just detected. Ramsay was keen to complete his research first. That afternoon, Dr Hampson arrived unexpectedly with some liquid air. This time, they used it to condense some of the 15 litres of argon containing the other noble gas impurities that they had isolated from the air. Then they trapped the first fractions that boiled off from this liquid. With more liquid air the following day, Saturday, they repeated the process, always trying to trap the most volatile portions evaporating from the liquefied argon. On Sunday, after the final purification, they came to view the spectrum of the most volatile fraction. Travers recalled: 'The leads from the induction coil were connected with the terminals of the vacuum tube, and as one of us turned on the current, we each picked up one of the little direct-vision spectroscopes which lay on the bench. But this time we had no need to use the prism to decide whether or not we were dealing with a new gas. The blaze of crimson light from the tube told its own story, and it was a sight to dwell upon and never to forget. It was worth the struggle of the previous two years; and all the difficulties yet to be overcome before the research was finished. The *undiscovered gas* had come to light in a manner which was no less than dramatic. For the moment the actual spectrum of the gas did not matter in the least, for nothing in the world gave a glow such as we had seen.'[46]

Nowadays this remarkable crimson glow of neon light is familiar to all of us through its use in advertisement signs. Technically, colours other than red should not be called 'neon' lights; these colours are obtained by the application of fluorescent coatings on the glass tubes, which glow with the desired hue when excited by ultraviolet or blue-violet light emitted from other gases in the tubes such as mercury or argon. But the bright red lights are entirely due to pure neon—when the current is switched off, the glass tube is completely clear and empty-looking. Perhaps if young Willie Ramsay had seen the light from the purified neon first, he might have thought of a different name for the new gas.

The Final Stranger

There was one more rare gas to be obtained from the air. One evening in mid-July, Ramsay and Travers were working late, separating some argon-krypton residues. They were just packing up to go home when Travers noticed a tiny bubble of gas was left in the pump because of the last gas that boiled off after the cold source was removed. It was most likely to be carbon dioxide, but Travers thought he might as well collect it anyway, even though it meant he then missed the last underground train home. The next day, he removed carbon dioxide using aqueous alkali to leave about 0.3 ml of gas, which he introduced into the tube to view its spectrum. He recorded in his note book: 'Krypton yellow appeared very faint, the green almost absent. Several red lines, three brilliant and equidistant, and several blue lines were seen. Is this pure krypton, at a pressure which does not bring out the yellow and green, or a new gas. Probably the latter!'[47] He later notes that the most striking feature was the beautiful blue glow from the tube: 'After a long search for a name suggesting the blue glow of the spectrum' they eventually settled on the name 'xenon', from the Greek for stranger. They thought 'all Greek and Latin roots indicating blue colour had long before been appropriated by organic chemists, and the name fixed upon had the merit that the symbol Xe was at least distinctive'.

Just like the original announcement of argon by Rayleigh and Ramsay, the discovery of the last of the rare gases found by Ramsay was revealed at a meeting of the British Association. William Crookes was the chair of the meeting, and he announced that he had also found a new element, which, just like the thallium he had found years before, he proposed to name after its spectrum— even though this time the spectrum was invisible to the eye. He writes, 'As the group of lines which betrayed its existence stand alone, almost at the extreme end of the ultra-violet spectrum, I propose to name the newest of the elements Monium, from the Greek μόνος [monos], alone.' He adds, 'Although caught by the searching rays of the spectrum, Monium offers a direct contrast to the recently discovered gaseous elements, by having a strongly marked individuality; but although so young and wilful, it is willing to enter into any number of chemical alliances.'[48] Crookes later changed the name of his new discovery to victorium, in honour of the golden jubilee of Queen Victoria (and, coincidentally, after he had received a knighthood from the monarch). Sadly, his discovery proved to be a false one; the substance was in fact a mixture of previously known rare earths. However, the summary of the chemistry section of the meeting suggests people had become rather tired of such new findings: 'The announcement of the discovery of two new elements, *Monium* and *Xenon*, must constitute a record for the first two days of the meeting, although new elements, especially amongst the rarer earths and gases, hardly excite the interest that similar discoveries did some years back.'[49]

The chemists had now discovered five new elements in the atmosphere that had been previously overlooked due to their inert properties and minute proportions. Although the inert gases remaining after the oxygen and nitrogen had been removed from air were noticed by Cavendish over a century earlier, chemists needed to be able to produce extremely low temperatures to separate significant quantities of the gases, and the spectroscope in order to detect and characterize them. To indicate just how tiny the fractions of the rare gases are in air, their percentage abundancies in dry air are given in Table 2, together with their boiling points.[50]

Xenon was the last of the rare gases that Ramsay was to discover, but not the last that he worked on. In what has been called his finest work, he studied the heaviest of the gaseous elements that can be found naturally occurring on Earth: radon. But in this region of the periodic table, things were beginning to fall apart.

Table 2. The abundancies of gases in dry air and their boiling points.

Gas	Fraction (%)	B.p.
nitrogen, N_2	78.084	77 K (−196° C)
oxygen, O_2	20.946	90 K (−183° C)
argon, Ar	0.934	87 K (−189° C)
neon, Ne	0.00182	27 K (−246° C)
helium, He	0.000524	4.2 K (−269° C)
krypton, Kr	0.000114	120 K (−153° C)
xenon, Xe	0.0000087	165 K (−108° C)

9

UNSTABLE ENDINGS

...the time has come for the systematizing chemist no longer to discriminate between natural and artificial elements...

—Paneth, 1947[1]

The Beginnings of Radioactivity

In 1896, Henri Becquerel (1852–1908) had discovered, by chance, the phenomenon of radioactivity, after he found that uranium salts left on top of covered photographic plates produced an image on the plates when they were later developed. Soon afterwards, thorium was also found to be radioactive. In 1898 Marie Curie (née Sklodovska) realized that certain minerals were more 'radioactive' (a term she first introduced) than could be rationalized by the amount of uranium or thorium that they contained. She guessed that they might contain trace amounts of an even more radioactive element, and during the long purification process, she eventually realized that two such elements were present. The naming of the first of these, discovered in July 1898, is described by her daughter Eve Curie in her biography of her mother:

> 'You will have to name it,' Pierre said to his young wife, in the same tone as if it were a question of choosing a name for little Irène [their first daughter]. The one-time Mlle Sklodovska reflected in silence for a moment. Then, her heart turning toward her own country which had been erased from the map of the world, she wondered vaguely if the scientific event would be published in Russia, Germany and Austria—the oppressor countries—and answered timidly: 'Could we call it "polonium"?'[2]

Marie Curie named the element after her homeland, Poland, but the country did not exist as a separate entity at that time, and her choice was something of a political statement.

The second element discovered by Marie and Pierre Curie was found to be millions of times more radioactive than uranium. This element they called 'radium' because of its intense radioactivity. Over three and a half years later, when they finally isolated a tenth of a gram of purified radium salts from tonnes of pitchblende ore, the Curies were delighted to find that the substance was spontaneously luminous.

Emanations

After the discovery that uranium and thorium were radioactive, in September 1899, Ernest Rutherford (1871–1937) made a further discovery: 'In addition to this ordinary radiation, I have found that thorium compounds continuously emit radio-active particles of some kind, which retain their radio-active powers for several minutes. This "emanation", as it will be termed for shortness, has the power of ionizing the gas in its neighbourhood and of passing through thin layers of metals, and, with great ease, through considerable thicknesses of paper.'[3] The following year, the German physicist Friedrich Ernst Dorn realized a similar 'emanation' was also emitted from the Curies' radium, but not directly from uranium.

Initially, it was not known exactly what these emanations were, but gradually, Rutherford and co-workers managed to show that 'the emanation is a chemically inert gas analogous in nature to the members of the argon family'.[4] Rutherford and Dorn had in fact found two isotopes of the same new element. Different isotopes of an element all have the same number of protons and electrons in the neutral atom, but differ in the number of neutrons in their nuclei. They have the same chemical properties, but differ slightly in their physical properties, especially those which depend on the masses of the atoms. The two isotopes of radon have very different half-lives (the time taken for half of the sample to undergo radioactive decay): for Dorn's emanation, it is 3.8 days, whereas for Rutherford's, it is just 55 seconds.

For a while, it was not appreciated that the emanations were even the same element. William Ramsay, who was already skilled in manipulating tiny quantities of gas, managed to isolate the longer-lived emanation of radium and record its spectrum. However, Ramsay did not like the name in use at the time. He wrote: 'The "emanation from radium", however, is a cumbrous expression, and sufficient evidence has now been accumulated that it is an element, accepting that word in the usual sense.' While he accepted that it was an element, he thought it was sufficiently different from all other elements to warrant a different type of name. He continues: 'Now, it appears advisable to devise a name which should recall its source, and, at the same time, by its termination, express the radical difference which undoubtedly exists between it and other elements. As it is derived from radium, why not name it simply "exradio"?'[5] He suggested similar terms could be applied to the emanations of other elements (actually now known to be other isotopes of radon): 'Should it be found that the emanation, which is supposed to be evolved from thorium, is really due to that element, and not to some other element mixed with thorium in exceedingly small amount, a similar name could be given, namely, "exthorio". If the existence of actinium as a definite element is established, its emanation would appropriately be named "exactinio". It is unlikely that others will be discovered, but, if they are, the same principle of nomenclature might be applied.'[6]

Ramsay's suggestions did not catch on. In 1910, when the nature of radioactivity was better understood, and it was appreciated that an atom of radium lost an alpha particle (a helium nucleus) to form an atom of the radium emanation, Ramsay made an accurate determination of its density to support this theory. At this point, he suggested a new name: 'The "emanation of radium" is a cumbrous name, and gives no indication of its position in the periodic table, a position which may now be taken as certain. To show its relation to gases of the argon series, it should receive a similar name; and the spectrum, the freezing-point, the boiling-point, the critical point, the density of the liquid, and the density of the gas, the last establishing, without doubt, the atomic weight of the element, having been determined in this laboratory, it only remains to give it a name. The name "niton", Nt, which has been used in this paper, is suggested as sufficiently distinctive.'[7] Ramsay derived the name from the Latin 'nitens', meaning 'shining', in recognition of its light-emitting properties. The literature became very confused with many different names applied to different isotopes of the element, including emanations of radium, thorium, and actinium and also the names radon, thoron, action, emanon, and niton. Eventually, the name radon was settled on in recognition of its longest-lived isotope, the emanation of radium; this was despite Marie Curie pushing for it to be named radioneon or radion.

Return of the Halogens

The element preceding radon in the periodic table, the fifth member of the halogens, was not discovered until 1940. Rather than being first found in nature, this element was actually prepared in a laboratory by bombarding a sample of bismuth with alpha particles. However, like other artificially synthesized elements, it did not immediately receive a name since it was hardly thought to count as a 'proper' element. Professor Friedrich Paneth, a chemist based at the University of Durham at the time, wrote in 1947:

> Almost five years ago, in a lecture to the Institute of Chemistry in London, the success that radioactive methods had achieved in the task of completing the Periodic System was described. In the table given, the place of element 87 was filled by the symbol of a newly discovered branch product in the actinium series; but in the places of elements 43, 61, 85 and 93 no symbols were inserted, although, as explained in some detail, atoms of all these four elements had been artificially produced.
>
> This denial of full citizenship to artificial elements seemed justified in those days. They had been produced in invisible amounts only, and they were unstable and usually not present on the earth; whereas in the case of all the natural elements, we could be sure that, even if they belonged to the radioactive families and were only represented by fairly short-lived isotopes, very considerable quantities always existed.[8]

In his paper, Paneth pleaded for the discoverers of the short-lived radioactive elements to give them names. The discoverers of the new halogen, Dale Corson, Kenneth MacKenzie, and Emilio Segrè, did just that:

> In 1940, we prepared the isotope of mass 211 of element 85 by bombarding bismuth with alpha particles accelerated in the 60-in cyclotron of the Radiation Laboratory of the University of California.
>
> At that time we established several chemical properties of element 85 and we made a fairly complete nuclear study of the isotope formed.
>
> It has been pointed out to us that a name should now be given to this new element, and following the system by which the lighter halogens, chlorine, bromine and iodine, have been named, namely, by modifying a Greek adjective denoting some property of the substance in question, we propose to call element 85 'astatine', from the Greek astatos, unstable. Astatine is, in fact, the only halogen without stable isotopes. The corresponding chemical symbol proposed is 'At'.[9]

The Superheavies

Since the synthesis of astatine, many more elements have been synthesized by momentarily fusing together the nuclei of two different elements to form the elements that come after uranium, the co-called transuranic elements, and more recently, the even more massive 'superheavy elements'. In addition to neptunium and plutonium which we met in Chapter 1, other elements were synthesized and named to recognize the work carried out at Berkeley. Famous scientists such as Albert Einstein, Pierre and Marie Curie, and Ernest Rutherford all have elements named after them. As, of course, does the creator of the Periodic Table, Dmitri Mendeleev. Currently his table is complete up to element 118, the last element of the seventh period (or row) in the table. In contrast to astatine's comparatively long half-life of just over seven hours, or radon's eternity of almost four days, many of the most recently synthesized elements have half-lives of mere fractions of seconds.

On 28 November 2016, the International Union of Pure and Applied Chemistry approved the names for the four most recently synthesized elements.

Element 113 was called nihonium, Nh, by the discoverers at RIKEN Nishina Center for Accelerator-Based Science in Japan. The name is derived from 'Nihon', which is one of two ways to say 'Japan' in Japanese; it literally means 'the land of the rising sun'. Element 115 was called moscovium, Mc, in recognition of the Moscow region that is home to the laboratories where it was first produced. Both of these two elements continued the tradition, started by Bergman and Berzelius, of using the termination '-ium' to distinguish metallic elements.

Element 117, at the bottom of the halogen group, was named tennessine. This name recognizes the contribution of the laboratories in the US state of Tennessee, who collaborated in the production of the most recent superheavy elements. However, even though it does not quite follow the group trend of 'modifying a Greek adjective denoting some property of the substance in question', it does still adopt the '-ine' termination first introduced by Humphry Davy with the naming of chlorine.

Finally, element 118, the final member of the periodic table (for now) and probably the last ever element of Group 18, the rare or 'noble' gases, has been named oganesson, Og. It is named after the Russian scientist Professor Yuri Oganessian (b. 1933), for his pioneering developments in the production of superheavy elements. Even though it is unlikely that any of its chemical properties will ever be investigated because of its extremely short half-life, it is still expected to be fairly unreactive, like the other members of its group, and so still keeps the ending '-on' following the trend set by Ramsay. Occasionally it has been suggested that the name for helium ought to be changed to helion, to fall in line with its relatives. Thankfully, this has not happened. To do so would lose some of its fascinating history.

NOTES

1. Ramsay, 'An Undiscovered Gas', 1897, p. 379.

CHAPTER 1

1. Ashmole, 1652, p. 313.
2. Sendivogius, 1650, p. 26.
3. Sendivogius, 1650, p. 19.
4. Webster, 1671, p. 42.
5. Lemery, *An Appendix to a Course of Chymistry*, 1680, pp. 21–2.
6. Webster, 1671, p. 42
7. Glaser, 1677, p. 55.
8. Salmon, 1671, p. 11.
9. Le Fèvre, *A Compleat Body of Chymistry*, 1664, Vol. II, p. 145.
10. Glaser, 1677, p. 65.
11. Schroeder, 1669, p. 180.
12. Lemery, *A Course of Chemistry*, 1686, pp. 70–3.
13. Glaser, 1677, p. 68.
14. Salmon, 1671, p. 11.
15. Schroeder, 1669, p. 187.
16. Glaser, 1677, p. 113.
17. Dover, [1733], pp. i–ii.
18. Salmon, 1671, p. 11.
19. Dover, [1733], pp. i–ii.
20. Charas, 1678, Book III, p. 175.
21. Maplet, 1567, fol. 7v.
22. Glaser, 1677, pp. 96–100.
23. Béguin, 1669, p. 113.
24. Goulard, 1769, p. 53.
25. Hesiod, 1988, p. 8.
26. Wall, 1783, p. 88.
27. Glaser, 1677, p. 90.
28. Boerhaave, *Elements of Chemistry*, 1735, Vol. 1, p. 26.
29. Hesiod, 1988, p. 8.
30. Salmon, 1671, p. 11.
31. Glaser, 1677, p. 105.
32. Glaser, 1677, p. 112.
33. Wall, 1783, p. 88.

34. Barba, 1674, pp. 89–91.
35. Klaproth, M., 1801, Vol. 1, p. 476.
36. Klaproth, 1791, p. 236.
37. Klaproth, M., 1801, Vol. 2 (1804), pp. 1–8.
38. Blackadder and Manderson, 1975.
39. Berzelius, *Jac. Berzelius Bref*, 1913, Vol. 1, tom. III, p. 162.
40. Pliny, 1601, Vol. 2, p. 629.

CHAPTER 2

1. Cronstedt, *An Essay towards a System of Mineralogy*, 1770, p. 241.
2. Webster, 1671, p. 330.
3. Webster, 1671, p. 335.
4. Webster, 1671, p. 335.
5. Bridge, 1994, p. 108.
6. Du Chesne, 1591, fol. 7v.
7. Webster, 1671, p. 335.
8. Glauber, *The Works*, 1689, Part II, p. 70.
9. Charas, 1678, Book III, p. 221.
10. Pepper, 1861, p. 438.
11. Scoffern, 1839, p. 186.
12. Waite, 1894, Vol. 1, p. 8.
13. Webster, 1671, p. 333.
14. Thomson, T., *A System of Chemistry*, 1802, Vol. 1, p. 182.
15. Geoffroy, 1736, pp. 189–90.
16. Bailey, *The Elder Pliny's Chapters on Chemical Subjects*, 1929, Vol. 1, p. 115.
17. Boerhaave, *A New Method of Chemistry*, 1727, pp. 66–7.
18. Schroeder, 1669, Book III, p. 220.
19. Valentinus, *The Last Will and Testament of Basil Valentine*, 1670, p. 233.
20. Valentinus, *The Triumphant Chariot of Antimony*, 1678, p. iv.
21. Valentinus, *The Triumphant Chariot of Antimony*, 1660, pp. 97–8.
22. Pomet, 1725, p. 358.
23. Geoffroy, 1736, pp. 189–90.
24. Glauber, *A Description of New Philosophical Furnaces*, 1651, p. 324.
25. Charas, 1678, p. 63.
26. Glauber, *A Description of New Philosophical Furnaces*, 1651, p. 324.
27. Lemery, *A Course of Chymistry*, 1677, p. 107.
28. Lemery, *A Course of Chymistry*, 1677, p. 109.
29. Pliny, 1601, Vol. 2, p. 469.
30. Bailey, *The Elder Pliny's Chapters on Chemical Subjects Part II*, 1932, p. 75.
31. Lemery, *A Course of Chymistry*, 1677, p. 134.
32. Geoffroy, 1736, p. 166.
33. Beckmann, 1797, Vol. 2, pp. 362–3.
34. Beckmann, 1797, Vol. 2, p. 363.
35. Agricola, *De natura fossilium*, 1955, footnote, p. 214.

36. Agricola, *De re metallica*, 1912, footnote, p. 217.

37. Heywood, 1635, pp. 568–9.

38. Bergman, 1784, Vol. 2, p. 232.

39. Watson, *Chemical Essays*, 1786, Vol. 4, p. 3.

40. Watson, *Chemical Essays*, 1786, Vol. 4, pp. 3–5.

41. Da Vigo, 'The Interpretacion of the Straunge Wordes', 1543.

42. Agricola, *De natura fossilium*, 1955, p. 96.

43. Needham, *Science and Civilisation in China*, 1974, Vol. 5 Part II, p. 212.

44. Boyle, *Essays of the Strange Subtilty...of Effluviums*, 1673, p. 19.

45. Stahl, 1730, p. 335.

46. Beckmann, 1797, Vol. 3, footnote, p. 93.

47. Agricola, *De re metallica*, 1912, footnote, p. 409.

CHAPTER 3

1. Woodall, 1617, p. 293.

2. Lydgate, 1498, fol. 3v.

3. Kircher, 1669, p. 35.

4. Kircher, 1669, p. 25.

5. Millar, 1754, p. 346.

6. Goethe, 1872, p. 363.

7. Pliny, 1601, Vol. 2, p. 557.

8. Lemery, *A Course of Chemistry*, 1698, p. 167.

9. Le Fèvre, *A Compleat Body of Chymistry*, 1664, Vol. 2, p. 335.

10. Helmont, 1662, p. 1154.

11. Ashmole, 1652, p. 365.

12. Gesner, 1576, fol. 186v.

13. Agricola, *De natura fossilium*, 1955, footnote, p. 213.

14. Bartholomaeus, 1582, fol. 265r.

15. Agricola, *De natura fossilium*, 1955, footnote, p. 47.

16. Agricola, *De natura fossilium*, 1955, footnote, p. 48.

17. Bailey, *The Elder Pliny's Chapters on Chemical Subjects, Part II*, 1932, p. 47.

18. Magnus, *Book of Minerals*, 1967, p. 243.

19. Gesner, 1576, fol. 195r.

20. Le Fèvre, *A Compleat Body of Chymistry*, 1664, p. 343.

21. Gesner, 1576, fol. 187v.

22. Musgrave, 1779, p. 21.

23. Frobenius, 'Spiritus Vini Aethereus, and the Phosphorus Urinae', 1733–1734, p. 55.

24. Ray, 1673, p. 235.

25. Ray, 1673, p. 236.

26. Principe, 2016, p. 130.

27. Principe, 2016.

28. Kunckel, 1716, p. 656.

29. Davis, T., 1927, p. 1106.

30. Davis, T., 1927, p. 1108.

31. Davis, T., 1927, p. 1109.
32. Davis, T., 1927, pp. 1110–11.
33. Davis, T., 1927, p. 1111.
34. Prandtl, 1948, pp. 416–17.
35. Prandtl, 1948, pp. 415–16.
36. Hooke, 1678, pp. 57–66.
37. Hooke, 1678, p. 61.
38. Boyle, *The Aerial Noctiluca*, 1680, p. 12.
39. Boyle, *The Aerial Noctiluca*, 1680, p. 5.
40. Boyle, *The Aerial Noctiluca*, 1680, p. 9.
41. Boyle, *The Aerial Noctiluca*, 1680, p. 105.
42. Krafft, 1969, pp. 665–7.
43. Boyle, *The Icy Noctiluca*, 1681/2, p. 19
44. Boyle, *The Icy Noctiluca*, 1681/2, p. 36.
45. Boyle, *The Icy Noctiluca*, 1681/2, pp. 78–9.
46. Y-Worth, *Chymicus Rationalis*, 1692, p. 50.
47. Frobenius and Hanckewitz, 'Experiments upon the Phosphorus', 1733–1734, p. 65.
48. Frobenius and Hanckewitz, 'Experiments upon the Phosphorus', 1733–1734, p. 66.

CHAPTER 4

1. Lavoisier, 'L'Eau n'est point une substance simple', 1784, pp. 473–4.
2. Boyle, *The Sceptical Chymist*, 1661, p. 21.
3. Waite, 1894, Vol. 1, p. 150.
4. Hapelius, 1606, p. 39.
5. Helmont, 1662, p. 1154.
6. Watson, Anecdotes, 1817, pp. 28–9.
7. Watson, *Chemical Essays*, 1781, Vol. 1, p. 167.
8. Cavallo, 1781, p. 244.
9. Cavallo, 1781, p. 576.
10. Lavoisier, *Essays Physical and Chemical*, 1776, p. 5.
11. Helmont, 1662, p. 69.
12. Dickson, 1796, p. 132.
13. Dickson, 1796, p. 133.
14. Helmont, 1662, p. 300.
15. Thomson, G., 1675, p. 198.
16. Thomson, G., 1675, fol. A8r.
17. Thomson, G., 1675, p. 153.
18. Thomson, G., 1675, fol. F5r mislabelled, p. 73.
19. Paracelsus, *Four Treatises*, 1941, p. 231.
20. Sendivogius, 1650, fol. Fff2v.
21. Paracelsus, *Four Treatises*, 1941, pp. 231–2.
22. Paracelsus, *Four Treatises*, 1941, p. 241.
23. Paracelsus, *Four Treatises*, 1941, p. 232.
24. Helmont, 1662, p. 96.
25. Black, 'Experiments upon Magnesia alba', 1756, p. 164.

26. Black, 'Experiments upon Magnesia alba', 1756, pp. 174–5.
27. Black, 'Experiments upon Magnesia alba', 1756, pp. 184–5.
28. Helmont, 1662, p. 427.
29. Boyle, Tracts, 1672, pp. 64–5.
30. Maud and Lowther, 1735, p. 283.
31. Cavendish, 'Three Papers', 1766, p. 141.
32. Cavendish, 'Three Papers', 1766, p. 145.
33. Boyle, *The Spring of the Air*, 1660, p. 364.
34. Hales, 1727, p. 178.
35. Scheele, *Air and Fire*, 1780, p. 3.
36. Scheele, *Air and Fire*, 1780, pp. 34–5.
37. Scheele, *Air and Fire*, 1780, p. 35.
38. Scheele, *Air and Fire*, 1780, p. 154.
39. Dobbin, 1935, p. 370.
40. Dobbin, 1935, p. 373.
41. Dobbin, 1935, p. 374.
42. Priestley, *Experiments and Observations*, 1774, Vol. 1, p. 178.
43. Priestley, 'Further Discoveries in Air', 1775, p. 387.
44. Priestley, 'Further Discoveries in Air', 1775, p. 388.
45. Priestley, *Experiments and Observations*, 1775, Vol. 2, p. 101.
46. Priestley, *Experiments and Observations*, 1775, Vol. 2, p. 102.
47. Priestley, *Experiments and Observations*, 1775, Vol. 2, p. 36.
48. Lavoisier, *Essays Physical and Chemical*, 1776, pp. 416–7.
49. Lavoisier, *Essays Physical and Chemical*, 1776, p. 417.
50. Lavoisier, *Effects on Atmospheric Air*, 1783, pp. v–vi.
51. Lavoisier, *Effects on Atmospheric Air*, 1783, pp. 97–8.
52. Lavoisier, *Effects on Atmospheric Air*, 1783, p. 98.
53. Lavoisier, *Effects on Atmospheric Air*, 1783, p. ix.
54. Priestley, 'Experiments Relating to Phlogiston', 1783, pp. 399–400.
55. Lavoisier, *Effects on Atmospheric Air*, 1783, p. x.
56. Cavendish, Experiments on Air, 1784, p. 119.
57. Cavendish, Experiments on Air, 1784, p. 137.
58. Watt, 1846, pp. 19–20.
59. Cavendish, 'Experiments on Air', 1784, pp. 134–5.
60. Lavoisier, 'L'Eau n'est point une substance simple', 1784, pp. 473–4.
61. Guyton de Morveau, 1788, p. 12.
62. Guyton de Morveau, 1788, p. 23.
63. Guyton de Morveau, 1788, pp. 23–4.
64. Guyton de Morveau, 1788, pp. 24–5.
65. Anon., 'Lettre...sur la nouvelle nomenclature', 1787, p. 423.
66. Lavoisier, *Elements of Chemistry*, 1790, footnote on, p. 89.
67. Lavoisier, *Elements of Chemistry*, 1793, footnote on, p. 142.
68. Dickson, 1796, p. 51.
69. Dickson, 1796, p. 106.
70. Dickson, 1796, p. 107.
71. Sage, 1800, footnote, p. 312.

72. Dickson, 1796, pp. 109.
73. De Arejula, 1788, p. 25.
74. Lavoisier, *Elements of Chemistry*, 1790, p. 53.
75. Guyton de Morveau, 1788, p. 26.
76. Dickson, 1796, p. 139.
77. Chaptal, 1791, p. xxxiv.
78. Chaptal, 1791, p. xxxv.
79. Beddoes and Watt, 1796, Part V, footnote, pp. 46–7.
80. Berkenhout, 1788, pp. xiii–xiv.
81. Berkenhout, 1788, p. xiv.
82. Black, *Elements of Chemistry*, 1803, Vol. 2, pp. 221–2.
83. Guyton de Morveau, 1788, pp. x–xi.
84. Davy, 'An Essay on Heat, Light', 1799, pp. 37–8.
85. Davy, 'Letter from Mr Davy', 1800, p. 517.

CHAPTER 5

1. Pliny, 1601, Vol. 2, p. 420.
2. Agricola, De re metallica, 1912, p. 558.
3. Helmont, 1662, p. 844.
4. Pliny, 1601, Vol. 2, p. 597.
5. Pliny, 1601, Vol. 2, p. 420.
6. Dodoens, 1578, pp. 115–16.
7. Agricola, De re metallica, 1912, p. 558.
8. Boerhaave, *Elements of Chemistry*, 1735, Vol. 1, p. 443.
9. Charas, 1678, Part III, p. 86.
10. Shakespeare, 1598, fol. B2v.
11. Needham, 1986, p. 99.
12. Needham, 1986, p. 97.
13. Needham, 1986, p. 117.
14. Biringuccio, 1942, p. 403.
15. Biringuccio, 1942, p. 111.
16. Whitehorne, 1562, fol. F3v–F4r unnumbered, and misnumbered as fol. 22.
17. Ercker, Fleta minor, 1683, p. 322.
18. Ercker, Ertzt unnd Bergkwercks Arten, 1580, fol. 125v.
19. Forchheimer, 1952.
20. Lemery, *A Course of Chemistry*, 1686, p. 289.
21. Wiegleb, 1789, pp. 192–3.
22. Barba, 1674, p. 29.
23. Dickson, 1796, pp. 239–40.
24. Pliny, 1601, Vol. 2, p. 415.
25. Charas, 1678, Part III, p. 139.
26. Ruscelli, 1578, fol. 70r.
27. Glauber, *A Description of New Philosophical Furnaces*, 1651, p. 162.
28. Priestley, *Experiments and Observations*, 1774, p. 166.
29. Guyton de Morveau, 1788, p. 49.

30. Tachenius, 1677, p. 9.
31. Dickson, 1796, pp. 239–40.
32. Guyton de Morveau, 1788, p. 49.
33. Guyton de Morveau, 1788, p. 49.
34. Nicholson, 1801, pp. 182–5.
35. Davy, 'The Decomposition of the Fixed Alkalies', 1808, p. 3.
36. Davy, 'The Decomposition of the Fixed Alkalies', 1808, p. 5.
37. Davy, *Collected Works*, 1839, Vol. 1, p. 109.
38. Davy, 'The Decomposition of the Fixed Alkalies', 1808, p. 5.
39. Davy, 'The Decomposition of the Fixed Alkalies', 1808, p. 13.
40. Davy, 'The Decomposition of the Fixed Alkalies', 1808, p. 31.
41. Davy, 'The Decomposition of the Fixed Alkalies', 1808, p. 32.
42. Davy, 'Chemischer Veränderungen', 1809, p. 157.
43. Berzelius, 'Essai sur la Nomenclature Chimique', 1811, p. 282, note 13.
44. Guyton de Morveau, 1788, p. 198.
45. Guyton de Morveau, 1788, p. 200.
46. Guyton de Morveau, 1788, p. 202.
47. Chenevix, 1802, pp. 230–1.
48. Gren, 1800, Vol. 2, pp. 465–6.
49. Berzelius, 'The Cause of Chemical Proportions', 1814, p. 51.
50. Berzelius, 'The Cause of Chemical Proportions', 1814, pp. 51–2.
51. Berzelius, *Jac. Berzelius bref. III Brefväxlingen mellan Berzelius och Thomas Thomson 1813–1825*, 1918, p. 18.
52. Berzelius, *Jac. Berzelius bref. III Brefväxlingen mellan Berzelius och Thomas Thomson 1813–1825*, 1918, pp. 24–5.

CHAPTER 6

1. Kirwan, *Elements of Mineralogy*, 1794, Vol. 1, p. 2.
2. Cronstedt, *An Essay towards a System of Mineralogy*, 1770, p. 8.
3. Scheffer, *Chemical Lectures*, 1992, p. 209.
4. Pettus, 1683, fol. D1r.
5. Drummond, 2007, pp. 55–7.
6. Klaproth, M., 1801, Vol. 2, footnote, p. 174.
7. Ward, 1640, p. 56.
8. Pliny, 1601, Vol. 2, p. 515.
9. Pliny, 1601, Vol. 2, p. 586.
10. Ward, 1640, p. 34.
11. Pliny, 1601, p. 587.
12. Magnus, *The Boke of Secretes*, 1560, fol. C1r.
13. Magnus, *The Boke of Secretes*, 1560, fol. C1r.
14. Magnus, *Book of Minerals*, 1967, p. 149.
15. Biringuccio, 1942, p. 115.
16. Pliny, 1601, Vol. 2, p. 597.
17. Helmont, 1662, p. 615.
18. Macquer, *A Dictionary of Chemistry*, 1777, Vol. 3, fol. S2r.

19. Haudicquer de Blancourt, 1699, p. 65.
20. Bergman, 1784, Vol. 2, p. 202.
21. Scheffer, *Chemiske föreläsningar*, 1775, p. 390; Scheffer, *Chemical Lectures*, 1992, p. 484.
22. Ashmole, 1652, p. 286.
23. Cronstedt, *Försök til mineralogie*, 1758, p. 106; Cronstedt, *An Essay towards a System of Mineralogy*, 1770, p. 123.
24. Hoffmann, *New Experiments and Observations upon Mineral Waters*, 1731, footnote, p. 78.
25. Hoffmann, *New Experiments and Observations upon Mineral Waters*, 1731, p. 75.
26. Black, *Experiments upon Magnesia Alba*, 1777, p. 4.
27. Pownall, 1825, p. 56.
28. Grew, 1697, p. 14.
29. Haudicquer de Blancourt, 1699, p. 65.
30. Ingram, 1767, p. 2.
31. Davis, T., 1927, p. 1106.
32. Domec, 1750, p. 17.
33. Guyton de Morveau, 1788, p. 79.
34. Chenevix, 1802, p. 84.
35. Dickson, 1796, p. 248.
36. Cronstedt, *An Essay towards a System of Mineralogy*, 1770, p. 18.
37. Pearson, 1794, p. 8.
38. Guyton de Morveau, 1788, p. 46.
39. Mathesius, 1562, fol. CXLr.
40. Albinus, 1589–1590, Vol. 2, p. 132.
41. Agricola, *De natura fossilium*, 1955, p. 96.
42. Ercker, *Fleta minor*, 1683, p. 308.
43. Henckel, 1756, Vol. 1, p. 130.
44. Minerophilo, 1730, p. 722.
45. Berzelius, 'Essai sur la Nomenclature Chimique', 1811, p. 280.
46. Agricola, *De natura fossilium*, 1955, p. 210.
47. Mathesius, 1562, fol. CXLVIr.
48. Camden, 1610, p. 676.
49. Webster, 1671, p. 280.
50. Pomet, 1725, Book 2, pp. 351–2.
51. Cramer, 1741, p. 181.
52. Scheele, *The Chemical Essays*, 1786, p. 227.
53. Scheele, *The Chemical Essays*, 1786, p. 242.
54. Scheele, *The Collected Papers of Carl Wilhelm Scheele*, 1931, p. 202.
55. Scheele, *The Chemical Essays*, 1786, p. 243.
56. Scheele, *The Collected Papers of Carl Wilhelm Scheele*, 1931, p. 206.
57. Lavoisier, *Elements of Chemistry*, 1790, p. 178.
58. Lavoisier, *Elements of Chemistry*, 1793, p. 219.
59. Tihavsky, 1791, pp. 283–4.
60. Lavoisier, *Elements of Chemistry*, 1793, p. 224.
61. Lavoisier, *Elements of Chemistry*, 1793, pp. 224–5.
62. Davy, *The Cecomposition of the Earths*, 1808, p. 346.
63. Davy, *Elements of Chemical Philosophy*, 1812, p. 350.

64. Clarke, 1817, p. 104.
65. Clarke, 1817, pp. 120–1.
66. Dolan, 1998, p. 157.
67. Davy, *The Decomposition of the Earths*, 1808, p. 353.
68. Klaproth, M., 1801, Vol. 1, p. 186.
69. Klaproth, M., 1801, Vol. 1, p. 199.
70. Thomson, T., *A System of Chemistry*, 1817, Vol. 1, p. 252.
71. Dixon and Wills, 1856, p. 507.
72. Dixon and Wills, 1856, p. 508.
73. Dixon and Wills, 1856, p. 510.

CHAPTER 7

1. Rider Meyer, *Real Fairy Folks: explorations in the world of atoms*, 1887.
2. Mrs Merrifield, 1849, Vol. 2, p. 592.
3. Cavendish, *Three Papers*, 1766, p. 158.
4. Priestley, *Experiments and Observations*, 1774, Vol. 1, pp. 147–8.
5. Lavoisier, *Elements of Chemistry*, 1790, p. 71.
6. Scheele, *The Collected Papers of Carl Wilhelm Scheele*, 1931, p. 91.
7. Fourcroy, 1788, Vol. 1, p. xliii.
8. Lavoisier, *Elements of Chemistry*, 1790, p. 72.
9. Davy, 'Researches on the Oxymuriatic Acid', 1810, p. 232.
10. Davy, 'Researches on the Oxymuriatic Acid', 1810, p. 236.
11. Davy, 'Researches on the Oxymuriatic Acid', 1810, p. 244.
12. Davy, 'Combinations of Oxymuriatic Gas and Oxygene', 1811, p. 32.
13. Mitchell, T. D., 1813, p. 106.
14. Mitchell, T. D., 1813, p. 106.
15. Davy, *Du gaz oximuriatique et de l'oxigène*, 1811, pp. 300–1.
16. Schweigger, 1811, p. 251.
17. Hofmann, 1882, p. 3141.
18. Agricola, *De natura fossilium*, 1955, footnote, p. 109.
19. Scheele, *The Chemical Essays*, 1786, p. 1.
20. Scheele, *The Chemical Essays*, 1786, p. 1.
21. Hill, 1774, p. 267.
22. Ampère, 'Lettres d'Ampère a Davy sur le Fluor', 1885, p. 9.
23. Ampère, 'D'une Classification naturelle pour Corps simples', 1816, pp. 22–4.
24. Davy, 'On a New Detonating Compound', 1813, p. 5.
25. Paris, 1831, Vol. 2, p. 18.
26. Davy, 'A New Substance which Becomes a Violet Coloured Gas', 1814, p. 74.
27. Davy, 'A New Substance which Becomes a Violet Coloured Gas', 1814, p. 91.
28. Ampère, 'D'une Classification naturelle pour Corps simples', 1816, pp. 22–4.
29. Thomson, T., *A System of Chemistry of Inorganic Bodies*, 1831, Vol. 1, footnote, p. 89.
30. Balard, 'Nouvelles des Science. Chimie', 1826, p. 376; Balard, *Muride, a Supposed New Elementary Substance*, 1826, p. 311.

31. Balard, 'Mémoire sur une Substance particulière contenue dans l'eau de la mer', 1826, p. 341; Balard, 'On a Peculiar Substance Contained in Sea Water', 1826, p. 383.
32. Casaseca, 1826, p. 527.
33. Turner, 1827, p. 695.
34. Gray, 1828, pp. 457–8.
35. Moissan, 1886, footnote, p. 1544.

CHAPTER 8

1. Travers, *The Discovery of the Rare Gases*, 1928, p. 23.
2. Kirchhoff and Bunsen, 'Chemische Analyse durch Spectralbeobachtungen', 1860, p. 186; Kirchhoff and Bunsen, 'Chemical Analysis by Spectrum-Observations', 1860, p. 107.
3. Kirchhoff and Bunsen, 'Chemische Analyse durch SpectralbeobachtungenZweite Abhandlung', 1861, p. 338; Kirchhoff and Bunsen, 'Chemical Analysis by Spectrum-Observations', 1861, p. 330.
4. Kirchhoff and Bunsen, 'Chemische Analyse durch SpectralbeobachtungenZweite Abhandlung', 1861, p. 338; Kirchhoff and Bunsen, 'Chemical Analysis by Spectrum-Observations', 1861, p. 330.
5. Kirchhoff and Bunsen, 'Chemische Analyse durch SpectralbeobachtungenZweite Abhandlung', 1861, p. 338; Kirchhoff and Bunsen, 'Chemical Analysis by Spectrum-Observations', 1861, p. 330.
6. Crookes, 'On the Existence of a New Element, Probably of the Sulphur Group', 1861, pp. 302 (*The Philosophical Magazine*) and 193 (*The Chemical News*).
7. Crookes, 'Further Remarks on the Supposed New Metalloid', 1861, p. 303.
8. Mendeleev, 1875, p. 293.
9. Mendeleev, 1875, p. 293.
10. Lecoq de Boisbaudran, 'Caractères chimiques et spectroscopiques d'un nouveau métal', 1875, p. 493; Lecoq de Boisbaudran, 'Chemical and Spectroscopic Characters of a New Metal, Gallium', 1875.
11. Anon., 'Bulletin', 1877.
12. Winkler, 'Discovery of a New Element', 1886, p. 418; Winkler, 'Germanium, Ge, ein neues, nicht-metallisches Element', p. 210.
13. Brunck, 1906, p. 4531.
14. Brunck, 1906, p. 4530.
15. Kirchhoff and Bunsen, 'Chemical Analysis by Spectrum-Observations', 1860, p. 108.
16. Kirchhoff and Bunsen, 'Chemical Analysis by Spectrum-Observations', 1860, p. 107.
17. Janssen, 'L'origine tellurique des raies de l'oxygene', 1889, p. 1035.
18. Janssen, 'L'éclipsé du mois d'août dernier', 1868.
19. Lockyer, 1869.
20. Thomson, W., 1871, pp. 267–8.
21. Carpenter, 1872, p. 308.
22. Mendeleeff, 1889, p. 643.
23. Rayleigh, 'Density of Nitrogen', 1892, p. 512.

24. Strutt, 1924, pp. 203–4.
25. Strutt, 1924, p. 204.
26. Strutt, 1924, p. 205.
27. The Times, 'Index', 1894, pp. 9 and 11.
28. Strutt, 1924, pp. 208–9.
29. Strutt, 1924, p. 211.
30. Strutt, 1924, p. 205.
31. Anon., 'Proceedings of Societies', 1894, p. 301.
32. Anon., 'Proceedings of Societies', 1894, p. 296.
33. Strutt, 1924, p. 219.
34. Rayleigh and Ramsay, 'Argon, a New Constituent of the Atmosphere', 1895, p. 270.
35. Travers, *A Life of Sir William Ramsay*, 1956, p. 110.
36. Ramsay, 'An Undiscovered Gas', 1897, p. 379.
37. Travers, *A Life of Sir William Ramsay*, 1956, pp. 133–4.
38. Hillebrand, 1890, p. 385.
39. Travers, *A Life of Sir William Ramsay*, 1956, p. 135.
40. Travers, *A Life of Sir William Ramsay*, 1956, p. 138.
41. Ramsay, 'An Undiscovered Gas', 1897, p. 380.
42. Travers, *A Life of Sir William Ramsay*, 1956, p. 173.
43. Travers, *A Life of Sir William Ramsay*, 1956, p. 174.
44. The Times, 'A New Gas', 1898, p. 5.
45. Travers, *A Life of Sir William Ramsay*, 1956, p. 175.
46. Travers, *The Discovery of the Rare Gases*, 1928, pp. 95–6.
47. Travers, *The Discovery of the Rare Gases*, 1928, p. 106.
48. Crookes, 'Inaugural Address of the British Association', 1898, p. 447.
49. Anon., 'Chemistry at the British Association', 1898, p. 556.
50. Brimblecombe, 1996, p. 3.

CHAPTER 9

1. Paneth, 1947, p. 8. Reprinted by permission from Macmillan Publishers Ltd: F. A. Paneth, 'The Making of the Missing Chemical Elements', *Nature* 159 (1947): 8–10, p. 8.
2. Curie, 1938, p. 169.
3. Rutherford, 'A Radio-active Substance Emitted from Thorium Compounds', 1900, p. 1.
4. Rutherford and Soddy, *The Cause and Nature of Radioactivity*, Part II, 1902, p. 581.
5. Ramsay and Collie, *The Spectrum of the Radium Emanation*, 1904, p. 476.
6. Ramsay and Collie, *The Spectrum of the Radium Emanation*, 1904, p. 476.
7. Whytlaw, Gray, and Ramsay, 1911, pp. 549–50.
8. Paneth, 1947, p. 8. Reprinted by permission from Macmillan Publishers Ltd: F. A. Paneth, 'The Making of the Missing Chemical Elements', *Nature* 159 (1947): 8–10, p. 8.
9. Corson, MacKenzie, and Segrè, 1947, p. 24. Reprinted by permission from Springer Nature: Nature, 'Astatine: The Element of Atomic Number 85', D. R. Corson et al., Copyright 1947.

FURTHER READING

There are a number of good books concerned with the elements and their discoveries.

Discovery of the Elements by Mary Elvira Weeks (*Journal of Chemical Education*, 1956). I have used the sixth edition. This is the standard work on the history of the discovery of the elements, and it includes many references to original sources.

Historical Studies in the Language of Chemistry by Maurice P. Crosland (Heinemann, 1962). An excellent scholarly work which looks at the part language has played in the development of chemistry, but it is not so focused on the elements themselves.

The Lost Elements: The Periodic Table's Shadow Side by Marco Fontani, Mariagrazia Costa, and Virginia Orna (Oxford University Press, 2015). An excellent, comprehensive text centred on the many names of elements that did not make it. It is largely concerned with discoveries from the nineteenth and twentieth centuries.

Nature's Building Blocks: An A–Z Guide to the Elements by John Emsely (Oxford University Press, 2001). An alphabetical guide to each of the elements, including pithy histories of each, but with no references.

A Tale of Seven Elements by Eric Scerri (Oxford University Press, 2013). A very detailed account of the discovery of the last seven elements that needed to be found to complete the periodic table up to uranium.

Superheavy: Making and Breaking the Periodic Table by Kit Chapman (Bloomsbury Publishing, 2019). A guide to the most recently synthesized and heaviest elements in the periodic table.

For anyone wishing to find more about the etymology of elements not covered in this book, an excellent starting point is Peter van der Krogt's website, *Elementymology & Elements Multidict* (http://www.vanderkrogt.net/elements/). At the time of writing, this site is regularly updated, and it includes many original references.

BIBLIOGRAPHY

Agricola, G. (1912). *De re metallica* (H. C. Hoover and L. H. Hoover, trans.). London: The Mining Magazine.

Agricola, G. (1955). *De natura fossilium (Textbook of Mineralogy)* (M. C. Bandy and J. A. Bandy, trans.). New York: Mineralogical Society of America.

Albinus, P. (1589–90). *Meisznische Land und Berg-Chronica, in welcher ein vollnstendige description des Landes, so zwischen der Elbe, Sala vnd Südödischen Behmischen gebirgen gelegen: So wol der dorinnen begriffenen auch anderer Bergwercken, sampt zugehörigen Metall un Metallar beschreibungen. Mit einuorleibten fürnehmen Sächsischen, Düringischen vnd Meissnischen Historien. Auch nicht wenig Tafeln, Wapen vnd Antiquiteten, derer etliche in Kupffer gestochen.* Dresden: Gimel Bergen.

Ampère, A.-M. (1816, May). 'D'une Classification naturelle pour Corps simples'. *Annales de Chimie et de Physique*, 2, 5–32.

Ampère, A.-M. (1885). 'Lettres d'Ampère a Davy sur le Fluor'. *Annales de Chimie et de Physique*, Series 6, vol. 4, 5–13.

Anon. (1787, December). 'Lettre aux auteurs du journal de physique, sur la nouvelle nomenclature chimique'. *Observations sur la physique, dur l'histoire naturelle et sur les arts*, 31, 418–24.

Anon. (1877, March 24). 'Bulletin'. *La Revue Politique et Littéarire revue des cours littéraires* (2E Série), 2E Serie Tome XII, p. 932.

Anon. (1894). 'Proceedings of Societies'. *The Chemical News and Journal of Physical Science*, 70, 301.

Anon. (1898). 'Chemistry at the British Association'. *Nature*, 58, 556.

Ashmole, E. (1652). *Theatrum Chemicum Britannicum*. London: Nath. Brooke.

Bailey, K. (1929). *The Elder Pliny's Chapters on Chemical Subjects*. London: Edward Arnold.

Bailey, K. (1932). *The Elder Pliny's Chapters on Chemical Subjects Part II*. London: Edward Arnold.

Balard, A. (1826). 'Memoir on a Peculiar Substance Contained in Sea Water'. *The Annals of Philosophy. New Series*, 12, 381–7, 411–26.

Balard, A. (1826). 'Mémoire sur une Substance particulière contenue dans l'eau de la mer'. *Annales de Chimie et de Physique*, 32, 337–84.

Balard, A. (1826). 'Nouvelles des Science. Chimie'. *Journal de Pharmacie et des Sciences Accessoires*, 12, 376–8.

Balard, A. (1826). 'Scientific Notices. Chemistry. Muride, a Supposed New Elementary Substance'. *The Annals of Philosophy. New Series*, 12, 311.

Barba, A. (1674). *The Art of Metals in Which Is Declared the Manner of Their Generation and the Concomitants of Them* (E. o. The R. H. Edward, trans.). London: S. Mearne.

Bartholomaeus, A. (1582). *Batman vppon Bartholome his booke De proprietatibus rerum, newly corrected, enlarged and amended: with such additions as are requisite, vnto euery seuerall booke: taken foorth of the most approued authors, the like heretofore not translated in English. Profitable for all estates, as well for the benefite of the mind as the bodie.* London.

Beckmann, J. (1797). *A History of Inventions and Discoveries* (W. Johnston, trans.). London: J. Bell.

Beddoes, T., and Watt, J. (1796). *Medical Cases and Speculations; Including Parts IV and V of Considerations on the Medicinal Powers, and the Production of Factitious Airs.* Bristol: J. Johnson.

Béguin, J. (1669). *Tyrocinium chymicum, or, Chymical Essays Acquired from the Fountain of Nature and Manual Experience.* London: Thomas Passenger.

Bergman, T. (1784). *Physical and Chemical Essays.* London: J. Murray.

Berkenhout, J. (1788). *First Lines of the Theory and Practice of Philosophical Chemistry.* London: T. Cadell.

Berzelius, J. J. (1811). 'Essai sur la Nomenclature Chimique' (J.-C. Delamétherie, ed.). *Journal de Physique, de Chimie et d'Histoire Naturelle*, 73, 253–86.

Berzelius, J. J. (1814). 'Essay on the Cause of Chemical Proportions, and on some Circumstances relating to them: together with a short and easy Method of expressing them' (T. Thomson, ed.). *Annals of Philosophy; or, Magazine of Chemistry, Mineralogy, Mechanics, Natural History, Agriculture, and the Arts*, 3, 443–54.

Berzelius, J. J. (1913). *Jac. Berzelius bref. III Brefväxlingen mellan Berzelius och Alexandre Marcet (1812–1822)* (H. G. Söderbaum, ed.). Uppsala.

Berzelius, J. J. (1918). *Jac. Berzelius bref. III Brefväxlingen mellan Berzelius och Thomas Thomson (1813–1825)* (Vol. VI). Uppsala.

Biringuccio, V. (1942). *The Pirotechnia of Vannoccio Biringuccio, translated from the Italian with an introduction and notes by Cyril Stanley Smith and Martha Teach Gnudi.* New York: The American Institute of Mining and Metallurgical Engineers.

Black, J. (1756). 'Art. VIII. Experiments upon Magnesia Alba, Quicklime, and Some Other Alcaline Substances'. *Essays and Observations, Physical and Literary. Read before a Society in Edinburgh, and Published by Them.* Volume II, 157–225.

Black, J. (1777). *Experiments upon Magnesia Alba, Quick-Lime, and Other Alcaline Substances.* Edinburgh: William Creech.

Black, J. (1803). *Lectures on the Elements of Chemistry, Delivered in the University of Edinburgh; by the late Joseph Black, M.D.* (J. Robison, ed.). London: Longman and Rees and Edinburgh: William Creech.

Blackadder, E. S., and Manderson, W. G. (1975). 'Occupational Absorption of Tellurium'. *British Journal of Industrial Medicine*, 32, 59–61.

Boerhaave, H. (1727). *A New Method of Chemistry; Including the Theory and Practice of That Art* (P. Shaw, trans.). London: J. Osborn and T. Longman.

Boerhaave, H. (1735). *Elements of Chemistry: Being the Annual Lectures of Herman Boerhaave, M.D.* (T. Dallowe, trans.). London: J. & J. Pemberton.

Boyle, R. (1660). *New experiments physico-mechanicall, touching the spring of the air, and its effects (made, for the most part, in a new pneumatical engine).* Oxford: Tho. Robinson.

Boyle, R. (1661). *The sceptical chymist: or Chymico-physical doubts and paradoxes, touching the spagyrist's principles commonly call'd hypostatical, as they are wont to be propos'd and defended by the generality of alchymists. Whereunto is praemis'd part of another discourse relating to the same subject.* London: J. Crooke.

Boyle, R. (1672). *Tracts written by the Honourable Robert Boyle Containing New experiments, touching the relation betwixt flame and air, and about explosions, an hydrostatical discourse occasion'd by some objections of Dr Henry More against some explications of new experiments made by the author of these tracts: to which is annex't, An hydrostatical letter, dilucidating an experiment about a way of weighing water in water, new experiments, of the positive or relative* levity of bodies under water, of the air's spring on bodies under water, about the differing pressure of heavy solids and fluids. London: Richard Davis.

Boyle, R. (1673). Essays of the strange subtilty great efficacy determinate nature of effluviums. To which are annext New experiments to make fire and flame ponderable: Together with A discovery of the perviousness of glass. London: M. Pitt.

Boyle, R. (1680). The aerial noctiluca: or Some new phoenomena, and a process of a factitious self-shining substance: Imparted in a letter to a friend, living in the country. London: Nath. Ranew.

Boyle, R. (1681/2). New experiments, and observations, made upon the icy noctiluca imparted in a letter to a friend living in the country: to which is annexed A chymical paradox. London: B. Tooke.

Bridge, D. (1994). 'The German Miners at Keswick and the Question of Bismuth'. Bulletin of the Peak District Mines Historical Society, 12, 108–12.

Brimblecombe, P. (1996). Air Composition and Chemistry (2nd ed.). Cambridge: Cambridge University Press.

Brunck, O. (1906). 'Obituary: Clemens Winkler'. Berichte der Deutschen Chemischen Gessellschaft, 39, pp. 4491–548.

Camden, W. (1610). Britain, or, A chorographicall description of the most flourishing kingdomes, England, Scotland, and Ireland, and the ilands adioyning, out of the depth of antiqvitie: beautified with mappes of the severall shires of England (P. Holland, trans.). London.

Carpenter, W. (1872). 'Inaugural Address of Dr William Carpenter, F.R.S., President'. Nature, 6, 306–12.

Casaseca, J. L. (1826). 'Note relative à la dénomination du brôme'. Journal de Pharmacie et des Sciences Accessoires, 12, 526–7.

Cavallo, T. (1781). A treatise on the nature and properties of air, and other permanently elastic fluids. To which is prefixed, an introduction to chymistry. London: Printed for the Author.

Cavendish, H. (1766). 'Three Papers, Containing Experiments on Factitious Air'. Philosophical Transactions, 56, 141–84.

Cavendish, H. (1784). 'Experiments on Air'. Philosophical Transactions of the Royal Society of London, 74, 119–53.

Chaptal, J. (1791). Elements of Chemistry. London: G. G. J. & J. Robinson.

Charas, M. (1678). The royal pharmacopoeea, galenical and chymical according to the practice of the most eminent and learned physitians of France: and publish'd with their several approbations; by Moses Charras, th Kings Chief operator in his royal garden of plants; faithfully Englished; illustrated with several copper plates. London: John Starkey.

Chenevix, R. (1802). Remarks upon Chemical Nomenclature, According to the Principles of the French Neologists. London: J. Bell.

Clarke, E. D. (1817). 'Account of some Experiments Made with Newman's Blow-pipe, by inflaming a highly condensed Mixture of the gaseous Constituents of Water; in a Letter to the Editor, from Edward Daniel Clarke, LL.D. Professor of Mineralogy in the University of Cambridge'. The Journal of Science and the Arts, 2(3), 104–23.

Corson, D. R., Mackenzie, K. R., and Segrè, E. (1947). 'Astatine: the Element of Atomic Number 85'. Nature, 159, 24.

Cramer, J. A. (1741). Elements of the art of assaying metals, tr. from the Lat., to which are added notes and observations. London: L. Davis and C. Reymers.

Cronstedt, A. F. (1758). Försök til mineralogie, eller mineral-rikets upställning. Stockholm.

Cronstedt, A. F. (1770). An Essay towards a System of Mineralogy: by Axel Fredric Cronstedt. Translated from the original Swedish, with notes, by Gustav von Engestrom. London: Edward and Charles Dilly.

Crookes, W. (1861). 'Further Remarks on the Supposed New Metalloid'. *The Chemical News*, 3(76), 303.

Crookes, W. (1861). 'On the Existence of a New Element, Probably of the Sulphur Group'. *The London, Edinburgh, and Dublin Philosophical Magazine and Journal of Science*, 21(140), 301–305.

Crookes, W. (1861). 'On the Existence of a New Element, Probably of the Sulphur Group'. *The Chemical News*, 3(69), 194.

Crookes, W. (1898). 'Inaugural Address of the British Association'. *Nature*, 58, 438–48.

Curie, E. (1938). *Madame Curie*. London: William Heinemann Ltd.

Davis, T. (1927, September). 'Kunckel and the Early History of Phosphorus'. *Journal of Chemical Education*, 4, 1105–13.

Davis, T. L. (1924). 'Neglected Evidence in the History of Phlogiston together with Observations on the Doctrine of Forms and the History of Alchemy'. *Annals of Medical History*, 6, 280–7.

Davy, H. (1799). 'An Essay on Heat, Light, and the Combinations of Light'. In T. Beddoes, *Contributions to Physical and Medical Knowledge, Principally from the West of England, collected by Thomas Beddoes, M. D.* (pp. 5–147). Bristol: T. N. Longman and O. Rees.

Davy, H. (1800). 'Letter from Mr Davy, Superintendent of the Pneumatic Institution, to Mr Nicholson, on the Nitrous Oxide, or Gaseous Oxide of Azote, on Certain Facts Relating to Heat and Light, and on the Discovery of the Decomposition of the Carbonate and Sulphate of Ammoniac' (W. Nicholson, ed.). *A Journal of Natural Philosophy, Chemistry, and the Arts*, 3, 515–18.

Davy, H. (1808). 'Electro-Chemical Researches, on the Decomposition of the Earths; with Observations on the Metals Obtained from the Alkaline Earths, and on the Amalgam Procured from Ammonia'. *Philosophical Transactions*, 98, 333–70.

Davy, H. (1808). 'The Bakerian Lecture: On Some New Phenomena of Chemical Changes Produced by Electricity, Particularly the Decomposition of the Fixed Alkalies, and the Exhibition of the New Substances Which Constitute Their Bases; and on the General Nature of Alkaline Bodies'. *Philosophical Transactions*, 98, 1–44.

Davy, H. (1809). 'Ueber einige neue Erscheinungen chemischer Veränderungen, welche durch die Electricität bewirkt werden…' (L. W. Gilbert, ed. and trans.). *Annalen der Physik, Neue Folge*, 113–76.

Davy, H. (1810). 'Researches on the Oxymuriatic Acid, Its Nature and Combinations; and on the Elements of the Muriatic Acid. With Some Experiments on Sulphur and Phosphorus, Made in the Laboratory of the Royal Institution'. *Philosophical Transactions*, 100, 231–57.

Davy, H. (1811). 'Expériences sur quelques combinaisons du gaz oximuriatique et de l'oxigène, et sur les rapports chimiques de ces principes avec les corps combustibles; par M. H. Davy'. *Annales de Chimie*, 68, 298–333.

Davy, H. (1811). 'The Bakerian Lecture: On Some of the Combinations of Oxymuriatic Gas and Oxygene, and on the Chemical Relations of These Principles, to Inflammable Bodies'. *Philosophical Transactions*, 101, 1–25.

Davy, H. (1812). *Elements of Chemical Philosophy*. London: J. Johnson & Co.

Davy, H. (1813). 'On a New Detonating Compound'. *Philosophical Transactions*, 103, 1–7.

Davy, H. (1814). 'Some Experiments and Observations on a New Substance which Becomes a Violet Coloured Gas by Heat'. *Philosophical Transactions*, 104, 74–93.

Davy, H. (1839). *The Collected Works of Sir Humphry Davy, Bart* (J. Davy, ed.). London: Smith, Elder & Co.

de Arejula, J. M. (1788). *Reflexiones sobre la nueva nomenclatura química propuesta por M. de Morveau, de la Academia de Ciencias de Dijon, y MM. Lavoisier, Berthollet, y de Fourcroy, de la Real Academica de Ciencias de Paris*. Madrid: Don Antonio de Sancha.

Dickson, S. (1796). *An Essay on Chemical Nomenclature...in which are Comprised Observations on the Same Subject, by Richard Kirwan*. London: J. Johnson.

Dixon, E. S., and Wills, W. H. (1856, 13 December). 'Aluminium'. *Household Words: A Weekly Journal. Conducted by Charles Dickens*, 14(351), 507–10.

Dobbin, L. (1935). 'Daniel Rutherford's Inaugural Dissertation'. *Journal of Chemical Education*, 12(8), 370–5.

Dodoens, R. (1578). *A niewe herball, or historie of plantes: wherin is contayned the whole discourse and perfect description of all sortes of herbes and plantes: their diuers & sundry kindes: their straunge figures, fashions, and shapes: their names, natures, operations, and vertues: and that not onely of those whiche are here growyng in this our countrie of Englande, but of all others also of forrayne realmes, commonly vsed in physicke*. London: Gerard Dewes.

Dolan, B. P. (1998). 'Blowpipes and Batteries: Humphry Davy, Edward Daniel Clarke, and Experimental Chemistry in Early Nineteenth-Century Britain'. *Ambix*, 45(3), 137–62.

Domec, A. (1750). *Dissertacion physico-chimica mecanico-medica, sobre las excelentes virtudes...y modo de obrar de la magnessia blanca...Y discurso physico-chimico, sobre el mejor methodo de elaborarla, para conseguirla mas virtuosa; por Don Joseph Belilla*. Saragossa: J. Fort.

Dover, T. (1733). *Encomium argenti vivi: a treatise upon the Use and Properties of quicksilver; or, The Natural, Chymical, and Physical History of that surprising Mineral...*London: Stephen Austin.

Drummond, P. (2007). *Scottish Hill Names: The Origin and Meaning of the Names of Scotland's Hills and Mountains* (2nd ed.). Scottish Mountaineering Trust.

du Chesne, J. (1591). *A breefe aunswere of Iosephus Quercetanus Armeniacus, Doctor of Phisick, to the exposition of Iacobus Aubertus Vindonis, concerning the original, and causes of mettalles Set foorth against chimists*. London.

Ercker, L. (1580). *Beschreibung allerfürnemisten mineralischen Ertzt unnd Bergkwercks Arten, wie dieselbigen, und eine jede in Sonderheit jrer Natur und Eigenschafft nach, auss alle Metaln probirt und im kleinen Fewer sollen versucht werden, mit Erklärung etlicher fürnemer nützlichen Schmeltszwerck, im grossen Fewer auch Scheidung Goldts, Silbers, und anderer Metaln, sampt einem Bericht dess Kupffersaigerns, Messing brennens, und Salpeter siedens, auch aller saltzigen Minerischen proben, und was denen allen anhengig, in fünff Bücher verfast, dessgleichen zuuorn niemals in Druck kommen* (2nd ed.). Frankfurt am Mayn.

Ercker, L. (1683). *Fleta minor, or, The laws of art and nature, in knowing, judging, assaying, fining, refining, and inlarging the bodies of confined metals in two parts: the first part contains assays of Lazarus Erckern, Chief Prover (or Assay-Master General of the Empire of Germany) in V. books, originally written by him in the Teutonick language, and now translated into English: the second contains essays on metalick words, alphabetically composed, as a dictionary, by Sir John Pettus...*London: Printed for the Author.

Forchheimer, P. (1952). 'The Etymology of Saltpeter'. *Modern Language Notes*, 67(2), 103–6.

Fourcroy, A.-F. (1788). *Elements of natural history, and of chemistry: being the second edition of the elementary lectures on those sciences, first published in 1782, and now greatly enlarged and improved, by the author, M. de Fourcroy...*London: G. G. J. & J. Robinson.

Frobenius, S. A. (1733–1734). 'An Account of the Experiments Shewn by Sigismund August Frobenius, M.D.F.R.S. at a Meeting of the Royal Society on November 18, 1731, with His Spiritus Vini Aethereus, and the Phosphorus Urinae, from the Minutes of That Day, by Cromwell Mortimer, M.D.' *Philosophical Transactions*, 38, 55–8.

Frobenius, S. A., and Hanckewitz, A. G. (1733–4). 'An Account of Some Experiments upon the Phosphorus Urinae, which May Serve as an Explanation to Those Shewn to the Royal Society by Dr Frobenius, on November 18, 1731, Together with Several Observations Tending to Explain the Nature of That Wonderful Chemical Production, by Mr Ambrose Godfrey Hanckewitz, F.R.S.' *Philosophical Transactions*, 38, 58–70.

Geoffroy, É. F. (1736). *A Treatise of the Fossil, Vegetable, and Animal Substances, That Are Made Use of in Physick*. London: W. Innys and R. Manby.

Gesner, K. (1576). *The newe iewell of health wherein is contayned the most excellent secretes of phisicke and philosophie, deuided into fower bookes. in the which are the best approued remedies for the diseases as well inwarde as outwarde, of all the partes of mans bodie: treating very amplye of all dystillations of waters, of oyles, balmes, quintessences, with the extraction of artificiall saltes, the vse and preparation of antimonie, and potable gold. Gathered out of the best and most approued authors, by that excellent doctor Gesnerus. Also the Pictures, and maner to make the vessels, furnaces, and other instrumentes therevnto belonging* (G. Baker, trans.). London.

Glaser, C. (1677). *The compleat chymist, or, A new treatise of chymistry teaching by a short and easy method all its most necessary preparations*. London: John Starkey.

Glauber, J. R. (1651). *A description of new philosophical furnaces, or, A new art of distilling: divided into five parts: whereunto is added a description Of the tincture of gold, or, the true aurum potabile: also the first part of the mineral work*. London: Tho. Williams.

Glauber, J. R. (1689). *The works of the highly experienced and famous chymist, John Rudolph Glauber containing great variety of choice secrets in medicine and alchymy, in the working of metallick mines, and the separation of metals: also various cheap and easie ways of making s*. London: Printed for the Author.

Goethe, J. W. (1872). *The Auto-Biography of Goethe. Truth and Poetry: From My Own Life* (J. Oxenford, trans.). London: Bell & Daldy.

Goulard, T. (1769). *A Treatise on the Effects and Various Preparations of Lead, Particularly of the Extract of Saturn, for Different Chirurgical Disorders*. London: P. Elmsly.

Gray, S. F. (1828). *The Operative Chemist; Being a Practical Display of the Arts and Manufactures which Depend upon Chemical Principles*. London: Hurst, Chance, & Co.

Gren, F. A. (1800). *Principles of Modern Chemistry, Systematically Arranged, by Dr Frederic Charles Gren*. London: T. Cadell.

Grew, N. (1697). *A Treatise of the Nature and Use of the Bitter Purging Salt Contain'd in Epsom, and Such Other Waters*. London.

Guyton de Morveau, L. B. (1788). *Method of chymical nomenclature, proposed by Messrs. De Morveau, Lavoisier, Bertholet, and De Fourcroy. To which is added, a new system of chymical characters, adapted to the nomenclature by Mess. Hassenfratz and Adet* (S. James, trans.). London: G. Kerasley.

Hales, S. (1727). *Vegetable staticks: or, an account of some statical experiments on the sap in vegetables: being an essay towards a natural history of vegetation*. London: W. & J. Innys.

Hapelius, N. N. (1606). *Nova Disquisitio De Helia Artista Theophrasteo In Qua De metallorum transformatione, adversus Hagelii & Pererii Jesuitarum opiniones evidenter & solide disseritur*. Marpurgi: Rodolphi Hutwelckeri.

Haudicquer de Blancourt, J. (1699). *The Art of Glass: Showing How to Make All Sorts of Glass, Crystal, & Enamel*. London: Dan Brown et al.

Helmont, J. B. (1662). *Oriatrike, or, Physick refined. The common errors therein refuted, and the whole art reformed & rectified: being a new rise and progress of phylosophy and medicine for the destruction of diseases and prolongation of life*. London: Lodowick Loyd.

Henckel, F. (1756). *Introduction à la minéralogie, ou, connoissance des eaux, des sucs terrestres, des sels, des terres, des pierres, des minéraux et des métaux, avec une description abrégée des opérations de métallurgie. Ouvrage posthume de J.F. Henckel publie sous le titre de Henckelius in mineralogiâ redivivus & traduit de l'Allemand*. Paris: Guillaume Cavelier.

Hesiod. (1988). *Theogony and Works and Days* (M. L. West, trans.). Oxford: Oxford University Press.

Heywood, T. (1635). *The hierarchie of the blessed angells*. London: Adam Islip.

Hill, J. (1774). 'Appendix I. Observations on the New-Discovered Swedish Acid; and on the Stone from which It Is Obtained'. In Theophrastus, *[Theophrastou tou Eresiou Peri ton lithon biblion]: Theophrastus's History of stones: with an English version, and critical and philosophical notes, including the modern history of the gems, &c. described by that author, and of many other of the native fossils; by John Hill; to which are added, two letters, one to Dr James Parsons, F.R.S. on the colours of the sapphire and turquoise, and the other, to Martin Folkes, upon the effects of different menstruums on copper; both tending to illustrate the doctrine of the gems being colored by metalline particles* (2nd ed.) (J. Hill, trans.), pp. 267–78. London: Printed for the Author.

Hillebrand, W. F. (1890). 'On the Occurrence of Nitrogen in Uraniite and on the Composition of Uraninite in General'. *American Journal of Science*, Series 3, 40, 384–94.

Hoffmann, F. (1731). *New experiments and observations upon mineral waters: directing their farther use for the preservation of health, and the cure of diseases* (1st ed.) (P. Shaw, trans.). London: J. Osborn and T. Longman.

Hoffmann, F. (1743). *New experiments and observations upon mineral waters, directing their farther use for the preservation of health, and the cure of diseases* (2nd ed.) (P. Shaw, trans.). London: T. Longman.

Hofmann, A. W. (1882). 'Zur Erinnerung an Friedrich Wöhler' (H. Wichelhaus, ed.). *Berichte der Deutschen Chemischen Gesellschaft*, 15(2), 3127–290.

Hooke, R. (1678). *Lectures and Collections*. London: The Royal Society.

Ingram, D. (1767). *An enquiry into the origin and nature of magnesia alba: and the properties of Epsom waters. Demonstrating, that magnesia made with those waters exceeds all others*. London: W. Owen.

Janssen, J. (1868). 'Indication de quelques-uns des résultats obtenus à Cocanada, pendant l'éclipsé du mois d'août dernier, et à la suite de cette éclipse'. *Comptes Rendus Hebdomadaires des Séances de l'Académie des Sciences*, 67, 838–9.

Janssen, J. (1889). 'SPECTROSCOPIE—Sur l'origine tellurique des raies de l'oxygène dans le spectre solaire'. *Comptes Rendus des Séances de l'Académie des Sciences*, 108, 1035–7.

King, C. W. (1866). *Antique Gems: Their Origin, Uses, and Value as Interpreters of Ancient History: and as Illustrative of Ancient Art: with Hints to Gem Collectors* (2nd ed.). London: John Murray.

Kircher, A. (1669). *The vulcano's, or, Burning and fire-vomiting mountains famous in the world, with their remarkables*. London: John Allen.

Kirchhoff, G., and Bunsen, K. (1860). 'Chemical Analysis by Spectrum-Observations'. *The London, Edinburgh, and Dublin Philosophical Magazine and Journal of Science*, 20(131), 88–109.

Kirchhoff, G., and Bunsen, R. (1860). 'Chemische Analyse durch Spectralbeobachtungen'. *Annalen der Physik und Chemie*, 110(6), 161–89.

Kirchhoff, G., and Bunsen, R. (1861). 'Chemical Analysis by Spectrum-Observations'. Second memoir. *The London, Edinburgh, and Dublin Philosophical Magazine and Journal of Science*, 22(148 and 150), 329–49 and 498–510.

Kirchhoff, G., and Bunsen, R. (1861). 'Chemische Analyse durch Spectralbeobachtungen. Zweite Abhandlung'. *Annalen der Physik und Chemie*, 113(7), 337–81.

Kirwan, R. (1784). *Elements of Mineralogy*. London: P. Elmsely.

Kirwan, R. (1794). *Elements of Mineralogy* (2nd ed.). London: P. Elmsly.

Klaproth. (1791). 'Crell's Chemical Journal; Giving an Account of the Latest Discoveries in Chemistry'. *Crell's Journal*, 1, 236.

Klaproth, M. (1801). *Analytical Essays towards Promoting the Chemical Knowledge of Mineral Substances*. London: T. Cadell.

Krafft, F. (1969). 'Phosphorus. From Elemental Light to Chemical Element'. *Angew. Chem. internat. Edit.*, 8(9), 660–71.

Kunckel, J. (1716). *Johann Kunckel von Löwensterns,Königl. Schwedischen Berg-Raths, und der Käyserl. Leopold. Societät Mit-Gliede, d. Hermes III. Collegium physico-chymicum experimentale: oder, Laboratorium Chymicum: in welchem Deutlich und gründlich von den wahren Principiis in der Natur und denen gewürckten Dingen so wohl über als in der Erden, als Vegetabilien, Animalien, Mineralien, Metallen, wie auch deren wahrhafften Generation Eigenschafften und Scheidung: nebst der Transmutation und Verbesserung der Metallen gehandelt wird....* Hamburg.

Lavoisier, A.-L. (1776). *Essays physical and chemical, by M. Lavoisier, member of the Royal Academy of Sciences at Paris, &c.* (T. Henry, trans.). London: Joseph Johnson.

Lavoisier, A.-L. (1783). *Essays, on the effects produced by various processes on atmospheric air with a particular view to an investigation of the constitution of the acids*. Warrington: J. Johnson.

Lavoisier, A.-L. (1784). 'MÉMOIRE Dans lequel on a pour objet de prouver que l'Eau n'est point une substance simple, un élément proprement dir, mais qu'elle est subceptible de décomposition & de recomposition'. *Histoire de l'Académie Royale des Sciences. Année M.DCCLXXXI*, 468–94.

Lavoisier, A.-L. (1790). *Elements of Chemistry, in a New Systematic Order, Containing All the Modern Discoveries*. Edinburgh: William Creech.

Lavoisier, A.-L. (1793). *Elements of Chemistry, in a New Systematic Order, Containing All the Modern Discoveries* (2nd ed.) (R. Keer, trans.). Edinburgh: William Creech.

Le Fèvre, N. (1664). *A compleat body of chymistry wherein is contained whatsoever is necessary for the attaining to the curious knowledge of this art*. London: Octavian Pulleyn Junior.

Lecoq de Boisbaudran, P. É. (1875). 'Caractères chimiques et spectroscopiques d'un nouveau métal, le Gallium, découvert dans une blende de la mine de Pierrefitte, vallée d'Argelès (Pyrénées)'. *Comptes rendus hebdomadaires des séances de l'Académie des sciences*, 81, 493–5.

Lecoq de Boisbaudran, P. É. (1875). 'Chemical and Spectroscopic Characters of a New Metal, Gallium, Discovered in the Blende of the Mine of Pierrefitte, in the Vally of Argeles, Pyrenees'. *The Chemical News*, 32, 159.

Lemery, N. (1677). *A Course of Chymistry* (1st ed.). London: Walter Kettilby.

Lemery, N. (1680). *An appendix to a course of chymistry being additional remarks to the former operations*. London: Walter Kettilby.

Lemery, N. (1686). *A course of chemistry containing an easie method of preparing those chymical medicins which are used in physick: with curious remarks and useful discourses upon each preparation, for the benefit of such who desire to be instructed in the knowledge of this art* (2nd ed.). London: Walter Kettilby.

Lemery, N. (1698). *A course of chemistry, containing an easie method of preparing those chymical medicins which are used in physick. with curious remarks and useful discourses upon each preparation, for the benefit of such as desire to be instructed in the knowledge of this art* (3rd ed.). London: Walter Kettilby.

Lockyer. (1869). 'Notice of an Observation of the Spectrum of a Solar Prominence'. *Proceedings of the Royal Society of London*, 17, 91–2.

Lydgate, J. (1498). *Hrre [sic] folowyth the interpretac[i]on of the names of goddis and goddesses of this treatyse folowynge as poetes wryte.*

Macquer, P. J. (1777). *A Dictionary of Chemistry* (2nd ed.). London: T. Cadell and P. Elmsly.

Magnus, A. (1560). *The boke of secretes of Albertus Magnus of the vertues of herbes, stones, and certayne beasts: also, a boke of the same author, of the marvaylous thinges of the world, and of certaine effectes caused of certaine beastes.*

Magnus, A. (1967). *Book of Minerals* (D. Wyckoff, trans.). Oxford: Clarendon Press.

Maplet, J. (1567). *A greene forest, or A naturall historie vvherein may bee seene first the most sufferaigne vertues in all the whole kinde of stones & mettals.* London.

Mathesius, J. (1562). *Sarepta oder Bergpostill sampt der Jochimssthalischen kurtzen Chroniken.* Nürnberg.

Maud, J., and Lowther, S. (1735). 'A chemical experiment by Mr John Maud, serving to illustrate the phoenomenon of the Inflammable Air shewn to the Royal Society by Sir James Lowther, Bart, as Described in Philosoph. Transact. numb. 429'. *Philosophical Transactions*, 39, 282–5.

Mendeleeff, D. (1889). 'The Periodic Law of the Chemical Elements'. *Journal of the Chemical Society, Transactions*, 55, 634–56.

Mendeleev, D. (1875). 'Remarks in Connection with the Discovery of Gallium'. *The Chemical News*, 32, 293–5.

Merrifield, M. P. (1849). *Original Treatises dating from the XIIth to XVIII Centuries. On the Arts of Painting in oil, miniature, mosaic, and on glass; of gilding, dyeing, and the preparation of colours and artificial gems; preceded by a general introduction; with translations, preface, and notes.* London: John Murray.

Millar, J. (1754). *A new course of chemistry: in which the theory and practice of that art are delivered in a familiar and intelligible manner.* London: D. Browne.

Minerophilo. (1730). *Neues und Curieuses Bergwercks-Lexicon, worinnen nicht nur alle und iede beym Bergwerck, Schmeltz-Hütten, Brenn-Hause, Saiger-Hütten, Blau-Farben-Mühlen, Hammerwercken u. vorkommende Benennungen, sondern auch derer Materien, Gefässe, Instrumenten und Arbeits-Arten Beschreibung enthalten, Alles nach dem gebraüchlichen Bergmännischen Stylo, so wohl aus eigener Erfahrung, als auch aus bewehrtesten Scribenten mit besondern Fleiss zusammen getragen Und in Alphabetische Ordnung zu sehr beqvehmen Nachschlagen gebracht, von Minerophilo, Freibergensi.* Chemniz.

Mitchell, T. D. (1813). 'On Muriatic and Oxy-Muriatic Acids, Combutions, &c. &c.' *Memoirs of the Columbian Chemical Society of Philadelphia*, 1, 102–17.

Moissan, H. (1886). 'Mémoires Présentés. Chimie. Action d'un courant électrique sur l'acide fluorhydrique anhydre'. *Comptes rendus hebdomadaires des séances de l'Académie des sciences*, 102, 1543–4.

Musgrave, S. (1779). *Gulstonian Lectures Read at the College of Physicians February 15, 16, and 17.* London.

Needham, J. (1974). *Science and Civilisation in China. Volume 5: Chemistry and Chemical Technology. Part 2: Spagyrical Discovery ad Invention: Magisteries of Gold and Immortality.* Cambridge: Cambridge University Press.

Needham, J. (1986). *Science and Civilisation in China. Volume 5: Chemistry and Chemical Technology. Part 7: Military Technology; the Gunpowder Epic.* Cambridge: Cambridge University Press.

Nicholson, W. (1801). 'Account of the New Electrical or Galvanic Apparatus of Sig. Alex. Volta, and Experiments Performed with the Same' (W. Nicholson, ed.). *A Journal of Natural Philosophy, Chemistry, and the Arts*, 4, 179–87.

Paneth, F. A. (1947). 'The Making of the Missing Chemical Elements'. *Nature*, 159, 8–10.

Paracelsus. (1941). *Four Treatises of Theophrastus von Hohenheim Called Paracelsus* (H. E. Sigerist, ed. C. L. Temkin, G. Rosen, G. Zilboorg, and H. E. Sigerist, trans.). Baltimore: The John Hopkins Press.

Paris, J. A. (1831). *The Life of Sir Humphry Davy, Bart. LL.D. Late president of the Royal Society, Foreign Associate of the Royal Institute of France, &c, &c, &c*. London: Henry Colburn and Richard Bentley.

Pearson, G. (1794). *A translation of the table of chemical nomenclature: proposed by De Guyton, formerly De Morveau, Lavoisier, Bertholet, and De Fourcroy; with additions and alterations: to which are prefixed an explanation of the terms, and some observations on the new system of chemistry*. London: J. Johnson.

Pepper, J. H. (n.d.). *The Playbook of Metals including personal narratives of visits to coal, lead, copper, and tin mines* (A. N. ed.). London: George Routledge and Sons.

Pettus, J. (1683). 'Fleta minor, Spagyrick Laws, the Second Part. Containing Essays on Metallick Words: Alphabetically Composed, as a Dictionary to Lazarus Erckern'. In *Fleta minor, or, The laws of art and nature, in knowing, judging, assaying, fining, refining, and inlarging the bodies of confined metals in two parts*...London: Printed for the Author.

Pliny. (1601). *The Historie of the World. Commonly Called, the Naural Historie of C. Plinius Secundus* (P. Holland, trans.). London: Adam Islip.

Pomet, P. (1725). *A compleat history of druggs* (2nd. ed.). London: R. & J. Bonwicke.

Porta, G. (1658). *Natural magick by John Baptista Porta, a Neapolitane: in twenty books: 1 Of the causes of wonderful things. 2 Of the generation of animals. 3 Of the production of new plants. 4 Of increasing houshold-stuff. 5 Of changing metals. 6 Of counterfeiting gold. 7 Of the wonders of the load-stone. 8 Of strange cures. 9 Of beautifying women. 10 Of distillation. 11 Of perfuming. 12 Of artificial fires. 13 Of tempering steel. 14 Of cookery. 15 Of fishing, fowling, hunting, &c. 16 Of invisible writing. 17 Of strange glasses. 18 Of statick experiments. 19 Of pneumatick experiments. 20 Of the Chaos.; Wherein are set forth all the riches and delights of the natural sciences.* London: Thomas Young.

Pownall, H. (1825). *Some particulars relating to the history of Epsom: compiled from the best authorities; containing a...description of the origin of horse racing, and of Epsom races, with an account of the mineral waters, and the two celebrated palaces of Durdans and Nonsuch, &c., &c. To which is added, an appendix, containing a botanical survey of the neighbourhood.* Epsom; London: W. Dorling; J. Hearne.

Prandtl, W. (1948, August). 'Some early publications on phosphorus'. *Journal of Chemical Education*, 25(8), 414–19.

Priestley, J. (1774). *Experiments and Observations on Different Kinds of Air*. London: J. Johnson.

Priestley, J. (1775). 'An Account of Further Discoveries in Air'. By the Rev. Joseph Priestley, LL.D. F.R.S. in letters to Sir John Pringle, Bart. P.R.S. and the Rev. Dr Price, F.R.S. *Philosophical Transactions*, 65, 384–94.

Priestley, J. (1775). *Experiments and Observations on Different Kinds of Air*. London: J. Johnson.

Priestley, J. (1783). 'Experiments Relating to Phlogiston, and the Seeming Conversion of Water into Air'. *Philosophical Transactions*, 73, 398–434.

Principe, L. M. (2016, May). 'Chymical Exotica in the Seventeenth Century, or, How to Make the Bologna Stone'. *Ambix*, 63(2), 118–44.

Ramsay, W. (1897). 'An Undiscovered Gas. Opening Address by Prof. Ramsay of the Chemistry Section of the Meeting of the British Association'. *Nature*, 56, 378–82.

Ramsay, W., and Collie, J. N. (1904). 'The Spectrum of the Radium Emanation'. *Proceedings of the Royal Society of London*, 73, 470–6.

Ray, J. (1673). *Observations topographical, moral, & physiological made in a journey through part of the low-countries, Germany, Italy, and France.* London: The Royal Society.

Rayleigh. (1892). 'Density of Nitrogen'. *Nature*, 46, 512–13.

Rayleigh, and Ramsay, W. (1895). 'Argon, a New Constituent of the Atmosphere'. *Proceedings of the Royal Society of London*, 57, 265–87.

Ruscelli, G. (1578). *The third and last part of the Secretes of the Reuerend Maister Alexis of Piemont, by him collected out of diuers excellent authors, with a necessary table in the ende, conteyning all the matters treated of in this present worke. Englished by William Ward.* London: John Wyght.

Rutherford, E. (1900). 'A Radio-active Substance Emitted from Thorium Compounds'. *The London, Edinburgh, and Dublin Philosophical Magazine and Journal of Science*, 49, 1–14.

Rutherford, E., and Soddy, F. (1902). 'The Cause and Nature of Radioactivity. Part II'. *The London, Edinburgh, and Dublin Philosophical Magazine and Journal of Science*, 4, 569–85.

Sage, B.-G. (1800). 'Sur la décomposition de l'acide nitreux fumant, par le moyen du charbon' (J.-C. Delamétherie, ed.). *Journal de Physique, de Chimie, d'Histoire Naturelle et des Arts*, 50, 310–12.

Salmon, W. (1671). *Synopsis medicinae, or, A compendium of astrological, Galenical, & chymical physick philosophically deduced from the principles of Hermes and Hippocrates, in three books: the first, laying down signs and rules how the disease may be known, the second, how to judge whether it be curable or not, or may end in life or death, the third, shewing the way of curing according to the precepts of Galen and Paracelsus...* London: Richard Jones.

Scheele, K. W. (1780). *Chemical Observations and Experiments on Air and Fire* (J. R. Forster, trans.). London: J. Johnson.

Scheele, K. W. (1786). *The Chemical Essays of Charles-William Scheele. Translated from the Transactions of the Academy of Sciences at Stockholm. With additions.* London: J. Murray.

Scheele, K. W. (1931). *The Collected Papers of Carl Wilhelm Scheele; translated from the Swedish and German Originals by Leonard Dobbin* (L. Dobbin, trans.). London: G. Bell.

Scheffer, H. T. (1775). *Herr H. T. Scheffers Chemiske föreläsningar, rörande salter, jordarter, vatten, fetmor, metaller och färgning, samlade, i ordning stälde och med anmärkningar utgifne.* Uppsala: Tryckte på Bokhandlaren M. Swederi: Bekostnad hos Joh. Edman.

Scheffer, H. T. (1992). *Chemical Lectures of H. T. Scheffer* (T. Bergman, ed., and J. A. Schufle, trans.). Dordrecht, Boston, London: Kluwer Academic Publishers.

Schroeder, J. (1669). *The compleat chymical dispensatory in five books treating of all sorts of metals, precious stones and minerals, of all vegetables and animals and things that are taken from them, as musk, civet, & c., how rightly to know them and how they are to be used in physick, with their several doses: the like work never extant before: being very proper for all merchants, druggists, chirurgions and apothecaries, and such ingenious persons as study physick or philosophy.* London: Richard Clavell.

Schweigger, J. (1811). 'Nachschreiben des Herausgebers, die neue Nomenclatur betreffend'. *Journal für Chemie und Physik*, 3, 249–67.

Scoffern, J. (1839). *Chemistry No Mystery; or a Lecturer's Bequest.* London: Harvery and Darton.

Sendivogius, M. (1650). *A New Light of Alchymie.* London.

Shakespeare, W. (1598). *The History of Henrie the Fourth; with the Battell at Shrewsburie, betweene the King and Lord Henry Percy, Surnamed Henrie Hotspur of the North. With the humorous conceits of Sir John Falstalffe.* London: Andrew Wise.

Stahl, G. E. (1730). *Philosophical Principles of Universal Chemistry: or, the foundation of a scientifical manner of inquiring into and preparing the natural and artificial bodies for the uses of life*. London: John Osborn and Thomas Longman.

Strutt, R. J. (1924). *Life of John William Strutt, Third Baron Rayleigh, O.M.* London: Edward Arnold & Co.

Tachenius, O. (1677). *Otto Tachenius his Hippocrates chymicus, which discovers the ancient foundations of the late viperine salt and his clavis thereunto*. London: Nathan Crouch.

Thomson, G. (1675). *Ortho-methodoz itro-chymike: or the direct method of curing chymically. Wherein is conteined the original matter, and principal agent of all natural bodies. also the efficient and material cause of diseases in general. their therapeutick way and means. I.* London: B. Billingley, and S. Crouch.

Thomson, T. (1802). *A System of Chemistry*. Edinburgh: Bell & Bradfute.

Thomson, T. (1817). *A System of Chemistry, in Four Volumes* (5th ed.). London: Baldwin, Cradock, and Joy.

Thomson, T. (1831). *A System of Chemistry of Inorganic Bodies*. London and Edinburgh: Baldwin & Cradock, London, and William Blackwood, Edinburgh.

Thomson, W. (1871). 'Inaugural Address of Sir William Thomson, LL.D., F.R.S., President'. *Nature*, 4, 262–70.

Tihavsk (1791). 'On the Metals Obtained from the Simple Earths'. *Crell's Chemical Journal; Giving an Account of the Latest Discoveries in Chemistry, with Extracts from Various Foreign Transactions: Translated from the German with Occasional Additions*, 1, 283–306.

The Times. (1894, 14 August). 'Index'. pp. 9 & 11.

The Times. (1898, 7 June). 'A New Gas'. p. 5.

Travers, M. (1928). *The Discovery of the Rare Gases*. London: Edward Arnold.

Travers, M. (1956). *A Life of Sir William Ramsay*. London: Edward Arnold.

Turner, E. (1827). *Elements of Chemistry: Including the Recent Discoveries and Doctrines of the Science*. Edinburgh: William Tait.

Valentinus, B. (1660). *The Triumphant Chariot of Antimony*. London: Thomas Bruster.

Valentinus, B. (1670). *The Last Will and Testament of Basil Valentine*. London: Edward Brewster.

Valentinus, B. (1678). *Basil Valentine His Triumphant Chariot of Antimony*. London: Dorman Newman.

Waite, A. E. (1894). *The Hermetic and Alchemical Writings of Aureolus Phillipus Theophrastus Bombast, of Hohenheim, called Paracelsus the Great*. London: James Elliott & Co.

Wall, M. (1783). *Dissertations on Select Subjects in Chemistry and Medicine by Martin Wall, M.D. Physician at Oxford, Public Reader of Chemistry in that University, and late Fellow of New College*. Oxford: D. Prince and J. Cooke.

Ward, S. (1640). *The wonders of the load-stone. Or, The load-stone newly reduc't into a divine and morall use*. London: Peter Cole.

Watson, R. (1781). *Chemical Essays. By R. Watson, D.D.F.R.S. and Regius Professor of Divinity in the University of Cambridge*. Cambridge.

Watson, R. (1786). *Chemical Essays. By R. Watson, D.D. F.R.S. and Regius Professor of Divinity in the University of Cambridge*. Cambridge.

Watson, R. (1817). *Anecdotes of the Life of Richard Watson, Bishop of Landaff; written by himself at different intervals, and revised in 1814*. London: T. Cadell and W. Davies.

Watt, J. (1846). *Correspondence of the Late James Watt on His Discoveries of the Theory of the Composition of Water. With a Letter from His Son* (J. P. Muirhead, ed.). London: John Murray.

Webster, J. (1671). *Metallographia, or, A History of Metals*. London: Walter Kettilby.

Whitehorne, P. (1562). *Certain waies for the orderyng of souldiers in batteles after divers fashion, with their maner of marchyng: And also fygures of certaine new plattes for fortificacion of Townes: And more over, howe to make Saltpeter, Gunpoulder, and divers sortes of fireworkes or wilde fyre, with other thynges apertaining to the warres. Gathered and set foorthe by Peter Whitehorne.* London: Nicolas Englande.

Whytlaw Gray, R., and Ramsay, W. (1911). 'The Density of Niton ("Radium Emanation") and the Disintegration Theory'. *Proceedings of the Royal Society*, 84, 536–50.

Wiegleb, J. C. (1789). *A General System of Chemistry, Theoretical and Practical. Digested and Arranged, with a Particular View to Its Application to the Arts* (C. R. Hopson, trans.). London: J. & J. Robinson.

Winkler, C. (1886). 'Discovery of a New Element'. *Nature*, 33, p. 418.

Winkler, C. (n.d.). 'Germanium, Ge, ein neues, nicht-metallisches Element'. *Berichte der Deutschen Chemischen Gesellschaft*, 19, pp. 210–11.

Woodall, J. (1617). *The surgions mate, or A treatise discovering faithfully and plainely the due contents of the surgions chest.* London: Laurence Lisle.

Y-Worth, W. (1692). *Chymicus rationalis, or, The fundamental grounds of the chymical art rationally stated and demonstrated by various examples in distillation, rectification, and exhaltation of vinor spirits, tinctures, oyls, salts, powers, and oleosums.* London: Thomas Salusbury.

INDEX

Note: For the benefit of digital users, indexed terms that span two pages (e.g., 52–3) may, on occasion, appear on only one of those pages.